T0215196

SUSTAINABLE INTENSIFICATION OF AGRICULTURE

Sustainable intensification (SI) has emerged in recent years as a powerful new conceptualisation of agricultural sustainability and has been widely adopted in policy circles and debates. It is defined as a process or system where yields are increased without adverse environmental impact and without the cultivation of more land.

Co-written by Jules Pretty, one of the pioneers of the concept and internationally known and respected authority on sustainable agriculture, this book sets out current thinking and debates around sustainable agriculture and intensification. It recognises that world population is increasing rapidly, so that yields must increase on finite land and other resources to maintain food security. It provides the first widely accessible overview of the concept of SI as an innovative approach to agriculture and as a key element in the transition to a green economy. It presents evidence from around the world to show how various innovations are improving yields, resilience and farm incomes, particularly for 'resource constrained' smallholders in developing countries, but also in the developed world. It shows how SI is a fundamental departure from previous models of agricultural intensification. It also highlights the particular role and potential of small-scale farmers and the fundamental importance of social and human capital in designing and spreading effective innovations.

Jules Pretty is Professor of Environment and Society at the University of Essex, UK. His sole-authored books include *The East Country* (2017), *The Edge of Extinction* (2014), *This Luminous Coast* (2011), *The Earth Only Endures* (2007), *Agri-Culture* (2002) and *Regenerating Agriculture* (1995). He is the Chief Editor of the *International Journal of Agricultural Sustainability*. He received the UK honour of OBE in 2006 for services to sustainable agriculture, and an honorary degree from Ohio State University in 2009. He has made many media contributions, including to BBC radio, on agriculture, food and sustainability.

Zareen Pervez Bharucha is a Research Fellow in the Global Sustainability Institute, Anglia Ruskin University, UK, where she leads the Global Risk and Resilience research theme. Her work addresses sustainability transitions, resilience and adaptation with a particular focus on rainfed farms in India. She is also co-editor and author of *Ecocultures: Blueprints for Sustainable Communities* (2015) and Deputy Editor for the *International Journal of Agricultural Sustainability*.

Other books in the Earthscan Food and Agriculture Series

The Commons, Plant Breeding and Agricultural Research
Challenges for Food Security and Agrobiodiversity
Edited by Fabien Girard and Christine Frison

Agri-environmental Governance as an Assemblage
Multiplicity, Power, and Transformation
Edited by Jérémie Forney, Chris Rosin and Hugh Campbell

Agricultural Development and Sustainable Intensification
Technology and Policy Challenges in the Face of Climate Change
Edited by Sekhar Udaya Nagothu

Environmental Justice and Soy Agribusiness
By Robert Hafner

Farmers' Cooperatives and Sustainable Food Systems in Europe
Raquel Ajates Gonzalez

The New Peasantries
Rural Development in Times of Globalization
(Second Edition)
Jan Douwe van der Ploeg

Sustainable Intensification of Agriculture
Greening the World's Food Economy
Jules Pretty and Zareen Pervez Bharucha

For further details please visit the series page on the Routledge website: http://www.routledge.com/books/series/ECEFA/

SUSTAINABLE INTENSIFICATION OF AGRICULTURE

Greening the World's Food Economy

Jules Pretty and Zareen Pervez Bharucha

Routledge
Taylor & Francis Group
LONDON AND NEW YORK

from Routledge

First published 2018
by Routledge
2 Park Square, Milton Park, Abingdon, Oxon OX14 4RN

and by Routledge
711 Third Avenue, New York, NY 10017

Routledge is an imprint of the Taylor & Francis Group, an informa business

British Library Cataloguing-in-Publication Data
A catalogue record for this book is available from the British Library

Library of Congress Cataloging-in-Publication Data
A catalog record has been requested for this book

ISBN: 978-1-138-19601-8 (hbk)
ISBN: 978-1-138-19602-5 (pbk)
ISBN: 978-1-138-63804-4 (ebk)

Typeset in Bembo
by Apex CoVantage, LLC

CONTENTS

1

IT COULD BE A WONDERFUL WORLD

Progress towards sustainable intensification

Think of this. Half of all children you know now, just saw walking to the local school, watched playing in a park, will live into the 22nd century. They will have adopted sufficient components of healthy living to see them live a hundred years. Their life journeys will take them past the point where world population will stabilise, then start to fall in some places, for the first time in human history. They will know a world where agriculture, the work of producing food, improves the natural capital of the planet rather than depleting it. This will happen in the temperate regions and tropics, across all continents, up hill and down dale.

Not possible? Perhaps utopian? Surely other pressing problems will intervene: political disturbance, climate change, pest and disease, drought or flood. Some of these may represent possible existential threats, many will result in greater temporary hunger and ill-health. But inexorably, year on year, the world's farmers will produce the food, fuel and fibre we need, from no more agricultural land. They will do it with responsibility and care for our environments and people. They will be part of redesigned food systems in which healthy food is grown with respect for nature, and distributed more evenly. There have been many agricultural revolutions across the last ten thousand years of human history. We are now amidst another – and it could be the most important.

In this book, we marshal the evidence. We look back to mid-last century when world population was four-tenths of that today, yet there were more people hungry and in poverty. We will show changes in productivity of land and other factors of production. We will note how the previous agricultural revolution also brought considerable harm to environments, and often to people's health. It did not seem possible, at the time, to conceive of a productive agriculture that did not trade off valuable services from the environment. You want food? Well, stop worrying about the birds and bees, the clean atmosphere and pristine waters, the diverse forests and boggy swamps. Losses are simply the price you must pay to eat. This was the narrative.

In the final couple of decades of the 20th century, evidence began to emerge that alternative approaches might work at scale. An innovation frontier was conceived and crossed. The word *sustainable* came into common use, though still seemed to many to be hopelessly optimistic. But farmers small and large, supported by researchers and extensionists, businesses, government and non-government agencies, experimented, organised, shared and learned. Agriculture became part of the knowledge economy. It was not a factory of grinding instruments, with predictable outcomes. It became performance.

When Thomas Kuhn (1962) wrote first of paradigms in the *Structure of Scientific Revolutions*, he showed how most science and practice fills in the spaces of an existing and well-defined pattern of ideas and theories. He called this normal science, firmly based on one or more past achievements. It is hard to change habituated practices; they seem from our personal experiences to be correct. Transitions are hard, often threatening. Physicist Max Planck wistfully observed that "a new scientific truth does not triumph by convincing its opponents and making them see the light, but rather because its opponents eventually die". It is not for us to say whether sustainability represents a new paradigm for agriculture, but it certainly means farmers doing new things, and scientists of all disciplines working on new technologies and practices to support them. New is risky, new is disruptive. New could also mean a better future for people and the planet.

But it will not be easy. In this book, we focus on the sustainable production of chiefly food crops. This is one part of the puzzle. There are many deep challenges associated with food production and consumption. Much today is wasted, lost to pests post-harvest, through cosmetic choices by retailers, left on plates or too long in a fridge. In the past generation or so, about 30 years, the proportion of people in affluent parts of the world who are overweight or obese has dramatically increased. In some industrialised countries and communities, more than a third of adults and a fifth of children are defined as clinically obese. Their futures do not look benign.

We do not fully address here the pull of consumption choices, though we do wrap up with observations on how redesigned agriculture can contribute to greener economies where all consumption patterns are very different to those of today. Our focus here is on the redesign of farming systems that can help shape those individual choices and behaviours. We feed more than 6 billion people well, yet the system is broken (Rockström *et al.*, 2009; Roisin *et al.*, 2012; Ehrlich and Ehrlich, 2013; IFPRI, 2016; IPES-Food, 2017). Our hope is that new world-building can begin.

Calls for a new type of farming are not recent. The desire for agriculture to produce more food without environmental harm, even make positive contributions to natural and social capital, has been reflected in calls for a wide range of different types of more sustainable agriculture: for a doubly green revolution, alternative agriculture, agroecological intensification, green food systems, greener revolutions, *agriculture durable* and evergreen agriculture. All centre on the proposition that agricultural and wild systems should no longer be conceived of as separate entities. All see positive synergies between planetary health and society, rather than zero-sum trade-offs. In light of the need for agriculture to contribute directly to the resolution of other global challenges, there have also been calls for nutrition-sensitive, climate-smart and low carbon agriculture. This is a great deal to ask of a single economic sector.

But agriculture is also unlike any other sector. Earlier models of intensification drew sharper distinctions between wild and farmed lands, between technology and nature, between intensive and extensive. This intensification was premised on the view that agriculture was an economic sector separated from the environment, emerging from the philosophical dominance of a Cartesian view of nature as machine. This led to an assumption of two opposed entities: people with constructed systems of food production, and the wild or even brutal nature out in the wider environment.

Compatibility of the terms *sustainable* and *intensification* was hinted at in the 1980s, and then first used together in an examination of the status and potential of African agriculture (Pretty, 1997). Until this point, *intensification* had become synonymous with types of high throughput agriculture characterised as causing harm whilst producing food. At the same time, *sustainable* was often seen as a term to be applied to all that could be good about agriculture. The combination of the terms was an attempt to indicate that desirable ends (more food, better environment) could be achieved by a variety of means. The term was further popularised by a number of key reports from learned societies, governments and multilateral bodies, including the United Nations.

We define Sustainable intensification (SI) as a process or system where yields are maintained or increased without adverse environmental impact and without the conversion of more land (Pretty, 2008; Pretty and Bharucha, 2014, 2015). The concept is thus relatively open, in that it does not articulate or privilege any particular vision of agricultural production. Though certain agronomic practices or packages have come to be frequently discussed, these are an indicative rather than a closed list of what constitutes sustainable intensification. Rather than being composed of a particular set of practices or technologies, sustainable intensification emphasises ends rather than means, and does not predetermine technologies, species mix or particular design components. Sustainable intensification can be distinguished from earlier conceptions of intensification because of an explicit emphasis on a wider set of environmental and health outcomes compared with just productivity enhancement.

Efficiency, substitution and redesign

In 1980s, Stuart Hill of the radical Hawkesbury College in Sydney, then of McGill University, developed a new concept of change in agricultural systems. This helps plot both steps towards new and more effective systems, and sets a scale for ambition.

Hill observed "there is something seriously wrong with a society that requires one to argue for sustainability", and suggested there were three critical phases and options in a transition to sustainable agricultural systems:

1   Efficiency
2   Substitution
3   Redesign.

We find this ESR progression framework helpful in understanding what we have achieved on a path towards sustainability in agricultural and food systems, and how the focus should now be on systemic redesign (Hill, 1985, 2014; MacRae et al., 1993; Lamine, 2011; Wright et al., 2011). Hill also distinguishes between deep (eco-design and redesign-based) and shallow (substitution-based systems). At the end of this book, we will conclude with observations about a new knowledge economy for agriculture, and the potential for world-building.

Step 1: Efficiency

The first step focuses on making best use of resources within existing system configurations. Why waste costly inputs or resources? Efficiency gains include targeting inputs of fertiliser and pesticide to focus impact, reduce use and cause less pollution and damage to natural capital and human health. The first progression thus draws from prophylactic, calendar-based and reactive approaches towards problem cure and, then, prevention. Precision agriculture is a further example, using global positioning system (GPS), robotics and drones to reduce both financial costs and environmental externalities. Better machine design can reduce the use of fossil fuels. In these ways, the unnecessary use of external inputs is avoided, saving on resources and farmers money.

These can be argued to be brilliant basics (Morgan, 1999): they should be done by all diligent farmers, but will probably not be much noticed when undertaken. They also do not result in system change.

Step 2: Substitution

This step focuses on the use of new technologies and practices to replace existing ones that may be less effective on both productivity and sustainability grounds. In sustainable intensification, inorganic inputs are often substituted by existing or revitalised ecosystem services. The development of new crop varieties and livestock breeds is an example of substitution replacing less efficient system components with new ones. Beetle banks substitute for insecticides; releases of biological control agents can also substitute for inputs. Hydroponics is an extreme example of substitution, where water-based architectural systems replace the use of soils. No- and zero-tillage systems substitute new forms of direct seeding and weed management for inversion tillage. Substitution implies an increasing intensification of resources, making better use of existing land, water and biodiversity, as well as technologies.

Substitution approaches can result in compellingly different systems on a considerable path towards sustainability arising out of systemic change.

Step 3: Redesign

This third step centres on the design of agroecosystems to deliver the optimum amount of ecosystem services to aid production whilst ensuring that agricultural production processes improve the ecosystems on which they depend. Redesign harnesses agroecological processes such as nutrient cycling, biological nitrogen fixation, allelopathy, predation and parasitism. The aim is to minimise the impacts of agroecosystem management on externalities such as greenhouse gas emissions, clean water, carbon sequestration, biodiversity, and dispersal of pests, pathogens and weeds. Redesign is a fundamentally social challenge, as there is a need to make productive use of human capital in the form of knowledge and capacity to adapt and innovate, and social capital to resolve common landscape-scale or system-wide problems (such as water, pest or soil management).

Redesign could be the game changer, setting agriculture on a journey that never ends, but with a clear sense of multiple targets and potentially wide social benefits. Hill (2014) does note a paradox, indicating why it has been so hard to achieve deep redesign: the more effective any efficiency and substitution initiatives are, the more likely they are to protect and perpetuate the design characteristics of the system that is the root cause of many problems.

The game changer

The notion of redesign as potential game changer is nonetheless important: it suggests there is no single solution to the productivity and sustainability challenges in agriculture. Systems will need to learn and develop, addressing new opportunities and challenges as they emerge. Thus sustainable intensification will become a paradigm for continuous learning, where means will differ temporally and spatially to achieve desired ends. Systems will emerge from localised social and ecological contexts, and possibly diverge.

This suggests the job is never done. Ecological and economic conditions change. This is particularly well illustrated by the challenges for integrated pest management. Pests, diseases and weeds evolve, new pests and diseases emerge (often because of pesticide overuse, sometimes from material transfers along trade routes), and pests and diseases are easily transported or are carried to new locations (often where natural enemies do not exist). New pests that have emerged in recent years include banana leaf roller (Nepal), invasive cassava mealybug (South East Asia), cucumber mosaic virus (Bangladesh), tomato yellow leaf curl virus (West Africa) and cassava mosaic virus and brown streak virus (Uganda).

The papaya mealybug is a native of Mexico, and is worryingly intrusive. It spread to the Caribbean in 1994, jumped to Pacific islands by 2002, and was reported in Indonesia, India and Sri Lanka by 2008 (Myrick *et al.*, 2014). In each new location, there was an absence of natural enemies. Parasitoids were collected in Puerto Rico and released in India and Sri Lanka in 2009–10, producing first year benefits to farmers of the order of US$300 million. The releases also prevented spread of the pest to northern India. But, papaya mealybug had by then spread to Thailand and the Philippines, and soon was discovered in Ghana. It then rushed 4,000 km along the coasts of West and Central Africa. The pest's preferred host is papaya, but it is highly polyphagous, feeding on 80 other species. Parasitoids were released in West Africa in 2013. In South East Asia, it has now jumped to mulberry, cassava, tomato and eggplant. Each geographic spread, each shift of host, requires new redesigns of agricultural systems.

We suggest redesigned and sustainable agroecosystems will have four features:

i   they will be multifunctional within landscapes and economies;
ii   they will jointly produce food and other goods for farmers and markets, while contributing to a range of valued public goods;
iii they will be diverse, synergistic and tailored to social–ecological contexts;
iv   they will have new configurations of social capital, comprising relations of trust embodied in social organisations, horizontal and vertical partnerships between institutions; and of human capital, comprising leadership, ingenuity, management skills and capacity to innovate.

There are many pathways towards agricultural sustainability, and no single configuration of technologies, inputs and ecological management is more likely to be widely applicable than another. For the past 15 years, we have been editors of the *International Journal of Agricultural Sustainability*, a leading peer-reviewed journal on the sustainability of agricultural systems worldwide. We have seen to publication some 350 novel scientific papers by the end of 2017. More is happening on the ground. There are 570 million farms worldwide, 90 per cent of which are run by individuals and families. Small farms occupy 12 per cent of world agricultural land, yet produce 70 per cent of the world's food (Lowder *et al.*, 2016). There is much evidence of transformations on these farms, as well as on the larger farms of industrialised countries.

Agricultural systems with high levels of social and human assets are able to innovate in the face of uncertainty and farmer-to-farmer learning has been shown to be particularly important in implementing the context-specific, knowledge-intensive and regenerative practices of sustainable intensification. We will need learning systems that include experimenting, designing and planning, and taking action (Bawden, 1998). This is an open or soft systems approach, where the system is the platform of learning. This can cause fundamental changes in worldviews, precisely what may now be required to ensure successful transitions towards sustainable and higher productivity in agricultural systems worldwide. In the end, there will be the need to improve values, not just systems of production (Hill, 2014).

2

TWENTY-FIRST CENTURY AGRICULTURE AND FOOD

Agriculture, nature and the environment

We can trace contemporary interest in agricultural sustainability to environmental concerns that became apparent in the 1950s and 1960s (Carson, 1962; Ward and Dubos, 1972). But concepts and practices about sustainability date much further back, at least to the surviving texts from ancient China, India, Greece and Rome (King, 1911; Cato, 1979; Li Wenhua, 2001; Pretty, 2003; Conway, 2012). The prominent Roman agricultural writers Cato, Varro and Columella spoke of agriculture as having two core features: *agri* and *cultura* (the fields and the culture). Cato, in the opening of *Di Agri Cultura*, written some 2,200 years ago, celebrated the high regard in which farmers were held: "when our ancestors ... would praise a worthy person their praise took this form: good husbandman, good farmer; one so praised was thought to have received the greatest commendation". He also observed how land longevity can affect us: "a good piece of land will please you more at each visit".

It is in China, though, that there exists the greatest and most continuous record of agriculture's close ties to environments and culture. Li Wenhua (2001) dates records of integrated crop, tree, livestock and fish farming to the Shang–West Zhou dynasties of 1600–800 BC. Later, Mensius in 400 BC drew attention to the importance of tenure arrangements if individuals were to invest in improving systems that reap later rewards: "If a family owns a certain piece of land with mulberry trees around it, a house for breeding silkworms, domesticated animals raised in its yard for meat, and crop fields cultivated and managed properly for cereals, it will be prosperous and will not suffer starvation."

In an early recognition of the need for the sustainable use of nature and its resources, Mensius went on to observe: "If the forests are timely felled, then an abundant supply of timber and firewood is ensured; if the fishing net with relatively big holes is timely cast into the pond, then there will be no shortage of fish and turtle for use." Still later, other treatises such as the collectively written *Li Shi Chun Qiu* (239 BC) and the *Qi Min Yao Shu* by Jia Sixia (AD 600) celebrated the value of agriculture to communities and economies, and showed how to sustain food production without causing damage to the environment. These included rotation methods and green manures for soil fertility, rules and norms for collective management of resources, the raising of fish in rice fields, the regular use of animal manures. Li Wenhua (2001) wrote that these presented: "a picture of a prosperous, diversified rural economy and a vivid sketch of pastoral peace". F.H. King's seminal study of Chinese and Japanese systems, *Farmers of Forty Centuries* (1911), documented a wide range of both productive and sustainable practices that had persisted for thousands of years.

Over time, with the building of surpluses and the development of diversified economies, agriculture came increasingly to be framed as an economic sector separated from nature and the environment. The philosophical dominance of a Cartesian view of nature-as-machine was built on long-standing monotheistic separations between humans and nature. It led to a gradual erosion of explicit connections to nature and the emergence of two entities: people with their constructed systems of production, and wild or brutal nature out in the environment.

During recent years, with growing concerns for sustainability, different typologies have been developed to categorise shades of shallow- to deep-green thinking (Naess, 1973; Worster, 1994). For some, there is a more fundamental schism: whether nature exists independently of humans, or whether it is part of a postmodern condition. There are long-standing dangers in these dualisms that separate humans from nature, suggesting that nature has boundaries, and can exist only in enclaves such as national parks, wildernesses, reserves, protected areas and zoos. Yet, the idea of untouched wilderness is also a myth (Gomez-Pompa and Kaus, 1992): natural systems we see today have mostly emerged from human shaping and interaction. There is never a clear line keeping out the wild from agricultural systems, nor indeed from human shaping of the wild (Bharucha and Pretty, 2010).

Tied in with a conceptual separation between the farmed and the wild is a common view that non-agricultural societies represent an earlier stage of cultural evolution, or even worse are the outcome of devolution (Barnard, 1999). Cultural evolutionary views supposed that societies progressed from hunter-gatherer to agricultural to industrial (Meggers, 1954; Lathrap, 1968). We now know the flaws in these perspectives (Kent, 1989; Kelly, 1995). The landmark *Man the Hunter* conference and book (Lee and DeVore, 1968) showed hunter-gatherers to be rich, knowledgeable, sophisticated and, above all, different from one another. There was no single stage of human development, just different adaptations to specific ecological and social circumstances.

It is now better accepted that cultures are adapted to localities, and thus are configured with a wide variety of land uses and livelihoods. Thus, foraging and farming systems across the world are "overlapping, interdependent, contemporaneous, coequal and complementary" (Sponsel, 1989). This suggests that rural societies might be better known as variants of cultivator-hunters or farmer-foragers: some horticulturalists move, some hunter-gatherers are sedentary (Szuter and Bayham, 1989; Vickers, 1989; Kelly, 1995). Some groups maintain gardens for cultivated food as well as to attract antelopes, monkeys and birds for hunting (Posey, 1985). Many apparently hunter-gatherer and forager cultures farm; many agricultural communities use hundreds if not thousands of non-domesticated species and resources (Bharucha and Pretty, 2010).

As culture and nature are bound together (Berkes, 2017; Pilgrim and Pretty, 2010; Boehm *et al.*, 2014), and the various forms of land use potentially complementary to one another, this suggests scope for consideration of agriculture as a food-producing system with a significant role in influencing and being influenced by environmental services (NRC, 2010; Foresight, 2011; NEA, 2011; FAO, 2016a; Bioversity International, 2016). Going beyond older divisions between land-sharing and land-sparing, there is an emerging recognition of interdependencies between agricultural and non-agricultural landscapes and the contribution of both to global social–ecological well-being.

The concept of planetary boundaries attempts to assess how the biosphere as a whole is performing in terms of its ability to support life. In 2009 a team of researchers led by Johan Rockström at the Stockholm Resilience Centre looked at planetary-scale processes, such as climate regulation and nutrient cycling, essential to life and complex societies. They then attempted to quantify the 'safe operating space' for each of these processes (Rockström *et al.*, 2009; Steffen *et al.*, 2015). Breached boundaries indicate that a particular process is no longer operating within the parameters that have so far been able to support the biosphere. Agriculture is a key driver of breached boundaries. Equally, redesigned agricultural systems will play a central role in repairing these breaches (Rockström *et al.*, 2017). Six important planetary boundaries operate in tight feedback loops with global agriculture and, to a broad degree, Rockström *et al.* (2017) suggest it is possible to quantify the scale of changes required to repair or maintain boundaries. For climate change, the challenge is perhaps the greatest. Agriculture is currently the world's largest single carbon source; it will need to become a sink. Land-use change, itself connected with climate, but also biodiversity, needs to be halted, so that forests are no longer cleared for farming. Agriculture uses most of the world's freshwater; water productivity must increase by 50 per cent by 2030. Biodiversity loss is another breached planetary boundary. Here, Rockström *et al.* propose the rather ambitious target of "zero loss of biodiversity in agricultural landscapes". Finally, reducing the influx of agricultural pollutants into the environment remains a priority, despite some progress towards managing agricultural waste and run-off.

The challenge, though, is great. In order to provide sufficient food for increasing human populations and their changing consumption patterns, some still indicate that agriculture will have to expand into non-cultivated lands. However, the competition for land from other human activities makes this an increasingly costly solution, particularly if protecting biodiversity and the public goods provided by natural ecosystems (e.g. carbon storage in forests or soils) is given priority (MEA,

2005; FAO, 2017a). Others suggest that yield increases must be achieved through redoubled efforts to repeat the approaches of the 1960s Green Revolution; or that agricultural systems should embrace only biotechnology or become solely organic. What is clear despite these differences in approach and philosophy is that more will need to be made of existing agricultural land (Tilman *et al.*, 2011; Smith, 2013; FAO, 2016b, 2017b).

Agriculture will, in short, have to be intensified. Traditionally agricultural intensification has been defined in three ways: (i) increasing yields per hectare; (ii) increasing cropping intensity (i.e. two or more crops) per unit of land or other inputs (water), or livestock intensity (e.g. faster maturing breeds); and (iii) changing land use from low value crops or commodities to those that receive higher market prices or have better nutritional content. The notion of intensification remains controversial to some, as recent successes in increasing food production per unit of resource have often also caused environmental harm and disruption to social systems. The evidence shows that sustainable intensification could both promote transitions towards greener and knowledge-intensive economies as well as benefit from progress in other sectors, while helping to heal breached planetary boundaries.

Agricultural revolutions

Early agricultural revolutions in industrialised countries focused on expansion of agricultural area to increase total food production. Such extensification was later followed by intensified use of resources on the same land. In both Europe and North America, wilder lands often used as commons came to be enclosed and privatised. In Britain during the 18th and early 19th centuries, some 2.75 million hectares (Mha) of common land were enclosed by 4,000 Acts of Parliament, comprising 1.82 Mha of open-field arable, and 0.93 Mha of what were then called wastes (areas of wild biodiversity). Reinforcing the enclosures of the wild, 50 different poaching offences were introduced with the penalty of death. Today, there are 18 Mha of agricultural land in the UK, of which 4 Mha are currently under arable farming; half a million hectares remain as commons.

These enclosures and expansions were accompanied by rapid innovation in agriculture in Europe and North America. Over a period of about 150 years, crop and livestock production in Britain increased three- to fourfold, as innovative technologies, such as the seed drill, novel crops such as turnips and legumes, fertilisation methods, rotation patterns, selective livestock breeding, drainage and irrigation were developed by farmers and spread to others through tours, open days, farmer groups and publications, and then adapted to local conditions by rigorous experimentation (Pretty, 1991). We were taught that this was *the* agricultural revolution.

But communities suffered. Critics of older systems of landscape design called them inefficient and unproductive. The dominant narrative painted commoners as idle, their landscapes in need of improvement by wealthy owners who could invest in technological improvement. Over time, then, agriculture came to be viewed not as a form of intense communion with nature, but as an economic sector separate from the environment. The result was dramatic transformations of landscapes. The poet John Clare wrote, "Inclosure came and trampled on the grave/Of labour's rights and left the poor a slave ... And birds and trees and flowers without a name/ All sighed when lawless law's enclosure came" (*The Mores*).

In modern, affluent economies, changing numbers of farmers and average farm size show how first extensification occurred, followed by intensification combined with changes in farm size. In the USA, farm numbers increased steadily from 1.5 million to over 6 million from 1860 to the 1920s, stabilised for 30 years, then fell rapidly since the 1950s to today's 2 million. Over the same period, average farm size remained stable for a century, around 60–80 hectares; but climbed from the 1950s to today's average of approximately 180 hectares (Figure 2.1). During the past 50 years, 4 million farms have disappeared in the USA. In France, 9 million farms in 1880 became just 1.5 million by the 1990s. In Japan, 6 million farmers in 1950 fell to 4 million by 2000. To advocates of economic progress and narrow measures of efficiency, these were predictable losses, inevitable if there was to be progress in increasing aggregate food production. Farmers increased their productivity, the inefficient were consolidated (an anodyne way of saying their farm businesses were closed), and the remaining farms were better able to compete on world markets. This exodus of small and medium-sized farmers continues across many sectors: in the UK, the number of dairy farms fell from 36,000 to 14,000 between 1995 and 2014; in the USA, 1,700 dairy farms were lost from the total 42,000 in 2016 alone. This farm consolidation brought undesirable social side-effects. Small farms continue to produce most of the world's food and are the primary source of livelihood for most of the world's agricultural labour. Communities also benefit from hosting a diversity of small farms. Seminal research by Walter Goldschmidt (1978 [1946])

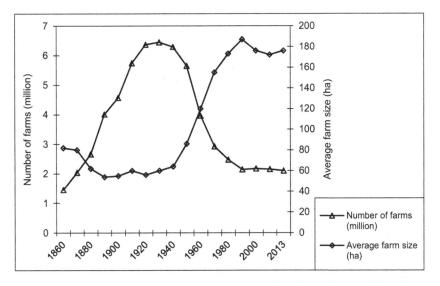

FIGURE 2.1   Number of farms and farm size in the USA, 1860–2013 (USDA, 2014)

on the two communities of Arvin and Dinuba in California's San Joaquin Valley, similar in all respects except for farm size, illustrated important social consequences of agricultural structures. Dinuba was characterised by small family farms, Arvin by large corporate enterprises. In Dinuba, Goldschmidt found a better quality of life, superior public services and facilities, more parks, shops and retail trade, twice the number of organisations for civic and social improvement, and better participation by the public. The small farm community was seen as a better place to live because, as Perelman (1976) later put it: "The small farm offered the opportunity for attachment to local culture and care for the surrounding land." Another study (Lobao, 1990) confirmed these findings: social connectedness, trust and participation in community life were greater where farm scale was smaller.

The mid-20th century then brought a new agricultural revolution, first in industrialised countries, and then in the tropics where it came to be known as the Green Revolution. New crop varieties and livestock breeds, combined with increased use of inorganic fertilisers, manufactured pesticides and machinery, together with better water control and increased field size, led to sharp increases in food production from agricultural systems. Many crop staples and livestock then showed remarkable productivity changes over a 50-year period (Figure 2.2; Table 2.1).

The result was a remarkable rise in food production, with increases across the world since good records began in the early 1960s (Figures 2.3 and 2.4). During the second half of the 20th century, intensification, rather than the spread of agricultural land, has been the prime driver of increased food production globally, and production has outpaced population growth. For each person today, there is 50 per

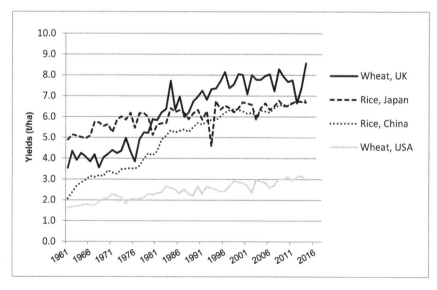

FIGURE 2.2 Profiles of yield changes for rice and wheat in four countries, 1961–2014 (FAOSTAT, 2017)

TABLE 2.1 Changing livestock productivity in the USA, 1955–2012

	1955	*2012*	*Ratio of productivity growth*
Beef cattle average live weight (kg per animal)	433	577	1.33
Pig average live weight (kg per animal)	108	124	1.15
Sheep average live weight (kg per animal)	43	62	1.44
Milk productivity (kg milk per dairy cow per year)	2,510	9,569	3.88
Broiler chicken weight (kg per bird)	1.39	2.61	1.88
Chicken layers (eggs per layer per year)	192	271	1.41

Source: USDA. 1955 and 2012. Agricultural Statistics. National Agricultural Statistics Service (NASS). At www.NASS.usda.gov/publications

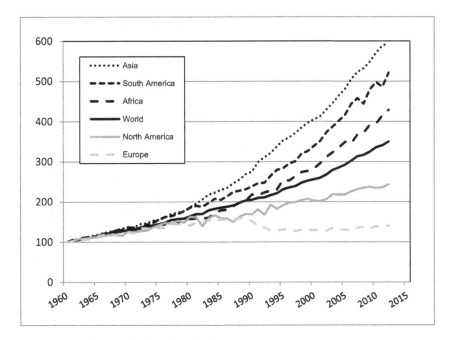

FIGURE 2.3   Global agricultural production (1961 = 100)

cent more food compared with each person in 1960. The growth in Asia and South America has been remarkable, Africa less so, with per capita growth only in patches, and then since the turn of the 21st century. The performance of different countries has been very different, China outpacing all countries, with significant per capita increases in Brazil, Indonesia and India.

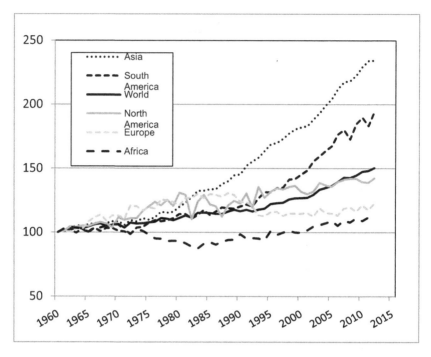

FIGURE 2.4   Global per capita agricultural production (1961 = 100) (FAOSTAT, 2017)

Though total agricultural area expanded by 11 per cent from 4.5 to 5 billion hectares and arable area from 1.27 to 1.4 billion hectares, aggregate world food production has more than tripled (with greatest growth in Asia and South America). The greatest increases have been in China, where an almost fivefold increase occurred, mostly during the 1980s–1990s. In industrialised countries, production started from a higher base; yet still it more than doubled in North America and grew by 40 per cent in Europe. Yet over the same time, this has been a period of rapid growth in world population: up 3 to more than 7 billion. The impact of the human footprint on the planet has further grown as consumption patterns also changed, converging particularly on those common in affluent countries (Pretty, 2013).

An important new challenge comes with shifting consumption patterns. Rising affluence is associated with nutrition transitions (Popkin, 1993), in which populations shift their consumption towards more saturated fats, sugars and refined foods. A key feature of the global dietary transition has been increased demand for animal protein. Livestock production has increased with a worldwide four-fold increase in numbers of chickens, doubling of pig numbers and 50 per cent increases in numbers of cattle, sheep and goats (Figure 2.5). One problem is that many of these animals are fed grain or pulses rather than being range-fed or grassland-raised. This introduces energetic inefficiencies into the world agricultural system.

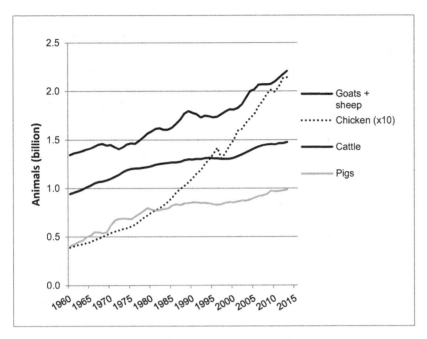

FIGURE 2.5 Global livestock numbers (FAOSTAT, 2017)

During the latter 20th century, the intensity of production on agricultural lands has also risen substantially: the area under irrigation and number of agricultural machines has grown by approximately twofold, fertiliser consumption by fourfold and nitrogen fertilisers by sevenfold. The use of synthetic pesticides has risen to 3.5 billion kg per year. Herbicides accounted for roughly half of use, insecticides and fungicides each about a quarter.

We now know this phase of agricultural intensification was accomplished at considerable expense to the environment. This in turn has made agricultural systems less efficient, by removing or degrading the environmental goods and services they need, such as groundwater for irrigation, pollinators and beneficial insects. These negative externalities begin to shift ideas about which agricultural systems were the most effective, suggesting again that new practices and systems that reduce negative externalities should be sought and developed.

It will not be easy

The scale of the food production and consumption challenges remains huge. By a narrow definition of calories per capita, global agriculture currently produces enough for all the world population to thrive. Yet, the world continues to face a continued triple burden of (i) under-nutrition (inadequate consumption of calories and protein), (ii) malnutrition (inadequate consumption of other important nutrients) and (iii) over-nutrition (excessive consumption of calories). This challenge is tightly enmeshed with other challenges of poverty, energy insecurity and breached planetary boundaries (Rockström *et al.*, 2009; Steffen *et al.*, 2015; FAO, 2017c).

The agricultural revolutions of the 20th century chiefly focused on reducing under-nutrition, seeking to boost the availability of calories through increased production of cereals and other staples. Progress has been good, but not good enough for millions of people (Figure 2.6). Today, some 800 million people worldwide remain undernourished, still one-ninth of all the people in the world population. Many countries failed to meet the Millennium Development Goal target of halving the number of hungry people over the period from 2000 to 2015; now all have until 2030 to meet the Sustainable Development Targets. The situation across the African continent remains particularly urgent. Of 34 countries requiring external food assistance in 2013, 27 were in Africa. Without significant effort, over 500 million people will still be food insecure in the region well into the 2020s.

In addition to chronic hunger, protracted food crises have become a norm. In the most vulnerable contexts, these may involve more than just crop failure or rising prices. The 2011 famine in Somalia touched on all four pillars of food security: there was a production shock, an access shock, a malnutrition crisis and increased instability of food sources (Maxwell, 2012). These shocks, like other disasters, stem from an increasingly complex combination of natural and human-made drivers. The 2007–08 food price spike, for instance, was caused by a combination of rising oil prices, market regulation and speculative activity. We can expect there to

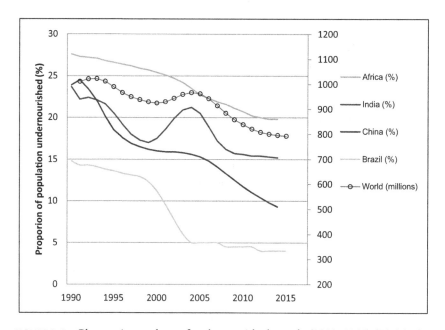

FIGURE 2.6 Changes in numbers of undernourished people (1990–2015) (FAOSTAT, 2017)

be many protracted crises arising from multiple causes, some with limited potential for immediate recovery (Foresight, 2011). In 2016, the numbers of chronically undernourished worldwide increased for the first time for a number of years, from 777 M in 2015 to 815 M. The main reason was a greater number of conflicts, often exacerbated by climate-related shocks (FAO, 2017b).

It is also thought that some 2 billion people suffer from various types of micronutrient malnutrition, deficiencies that particularly affect the health and human potential of women and children. Vitamin A deficiency is a public health concern in 80 countries, affecting over 7 million pregnant women and 127 million preschool children worldwide. Micronutrient deficiencies during critical life stages can have lasting impact on both individuals and their societies.

Third, over-nutrition now negatively affects the health of 600 million adults worldwide (FAO, 2017c). In the mid-1980s, average adult incidence of obesity in the UK was 3 per cent and in the USA 6 per cent. It then sharply rose to 24 per cent of adults in the UK, and to 38 per cent in the USA by 2015 (OECD, 2017). Now many wealthier groups in fast developing countries, particularly in richer urban settings, are making a similar transition to overweight and obese populations. Mexico is now the highest consumer of sugar-rich soft drinks, and has such a rapid growth in incidence of obesity that it has the second highest proportion of obese people in the world (Carolan, 2013; OECD, 2017). In India, diabetes and hypertension have emerged as major public health concerns (Shetty, 2012). Across Africa, despite the continued prevalence of under-nutrition, the incidence of obesity and diabetes is increasing (Frayne et al., 2014; Peer et al., 2014). By 2008, there were more obese people in the developing world than in rich countries, with numbers more than tripling since the 1980s (Stevens et al., 2012).

Thus what is important is not just increasing yields or producing more calories per capita. Globally, despite pockets of high dietary diversity amongst some land-based communities (Bharucha and Pretty, 2010), just 12 species contribute 80 per cent of dietary intake and global agriculture has come to focus on just 150 commercialised species. A narrow focus on calories and commercial staples has resulted in some unintended nutritional outcomes. Across south Asia, for example, cereal production increased by fourfold from 1970, yet this was achieved alongside a decline of 20 per cent in the cultivation of pulses. Thus, future efforts to intensify agriculture substantially will have to be nutrition-sensitive too. This implies focused attention on a wider range of nutritionally dense crops, including fruit, vegetables, legumes, animal products and non-cultivated species.

Improving agricultural growth is also imperative for reducing poverty. Agricultural growth is more effective at reducing poverty than general economic growth, which is especially important in countries with largely agrarian populations and suffering from high levels of poverty.

It is clear, however, that agricultural systems alone cannot solve world problems: conflicts have reduced agricultural production. Food production in 13 war-affected countries of sub-Saharan Africa over 1970–1994 was 12 per cent lower in war years compared with peace-adjusted values. Over the period 1970–1997, FAO (2000) estimated that conflict-related losses of agricultural outputs amounted to just over

US$120 billion ($4.3 billion per year, at 1995 prices). Agriculture can go some way towards addressing major threats such as climate change and biodiversity loss, but is itself made more vulnerable because of these. And, finally, agriculture alone cannot address global hunger, inequality and poverty. Instead, it can act as a catalyst for wider change, but momentum will have to be maintained beyond agriculture by committed policy action across economies.

The birds and the bees

Agriculture depends on biodiversity on and around farms, and diverse farming systems are essential for nutritional security. Yet, too often across the 20th century, intensification has meant simplification of agricultural landscapes, producing less diverse cropping patterns, narrow selections of crop traits, and monocultures. Agricultural externalities have also caused significant harm to off-farm biodiversity. This in turn jeopardises several nutritionally important crops. An astonishing 62 per cent of all IUCN globally threatened species are affected by agricultural systems (Maxwell *et al.*, 2016), making it imperative that we redesign relations between the farmed and the wild, and make peace with biodiversity on farms. This will have practical implications for food security too, not just through increased yields but by providing greater flows of non-cultivated but nutritionally important foods. Wild foods are critically important sources of protein, energy and micronutrients for those most vulnerable to hunger globally (Bharucha and Pretty, 2010; IUCN, 2013).

The majority of the global supply of vitamins A and C, folic acid, calcium, fluoride and iron comes from animal- and insect-pollinated crops (Eilers *et al.*, 2011). In the USA, bees contribute 11 per cent of agricultural gross domestic product (GDP); some 20 per cent of this is contributed by wild pollinators. Yet crop pollinators have been under long-term decline across parts of Europe and North America; managed honey bee populations have also been in decline. Introduced parasites, antibiotic-resistant pathogens, pesticide-resistant mites, encroachment of Africanised honey bees, and some new harmful pesticide compounds have all combined to reduce populations. At the same time, pollinator services are increasingly important, with rising demand for insect-pollinated crops and trees (Koh *et al.*, 2016).

In Europe, wild bird populations have been in sharp decline, with common farmland birds declining in numbers by 57 per cent between 1980 and 2013 (European Bird Census Council, 2015). Changes to cultivation practices are the most frequently reported pressure, in particular the shift from spring to winter cereals from the early 1980s. The relationship between agricultural intensification and the decline of farmland birds in England and Wales has been illustrated by Chamberlain *et al.* (2000), who concluded that multiple, interacting changes involved in intensification had caused declines in bird populations across a range of ecological niches.

In the UK, more than nine-tenths of flower-rich wild meadows have been lost since the 1940s, as well as half of heathland, lowland fens and valley and basin mires, and up to half of all ancient lowland woods and hedgerows. Hedgerows host a large number of threatened or rare species. Between 1998 and 2007, some 26,000 kilometres of managed hedgerows were lost, though now hedgerow removal is no longer permitted. Most of these changes to important biodiversity have been creeping, persistent and long term. Reversing them will require radical shifts in our relationship with the land.

The pesticide treadmill

Pathogens, weeds and invertebrates have always caused significant agricultural losses worldwide, and thus of course present a barrier to the achievement of global food security. Pesticide compounds have also long been used to control pests and diseases in agriculture (Carson, 1962; Conway and Pretty, 1991; Pretty, 2005; Zhang *et al.*, 2011). In 2500 BC, Sumerians used sulphur compounds for insect control. Later, seeds were treated by Chinese farmers with natural organic substances to protect against insects, mice and birds, while inorganic mercury and arsenic compounds were used to control body lice. A variety of fumigants, oil sprays and sulphur ointments were used in Greece and Rome, and Pliny recommended the use of arsenic as an insecticide. The widespread use of natural pesticides began in the 17th and 18th centuries in Europe: first nicotine and mercuric oxide, then copper sulphate as a fungicide in the early 1800s. By the mid-19th century, rotenone from derris and pyrethrum from chrysanthemum had been discovered, these were accompanied by rapid growth in the use of inorganic compounds, particularly of arsenic. Paris Green (copper arsenite) was first used in agriculture in 1867 (having been found to be a good killer of rats in Paris sewers), then Bordeaux mixture (copper sulphate and lime) was found to be effective against powdery mildew in 1882.

Now came phases of compounds and products, characterised at first by great efficacy then rapid emergence of evidence of harm. The early part of the 20th century was characterised by the increased use of dangerous compounds of arsenic, cyanide and mercury. Most were broad-acting in their effect on pests and diseases. Some others, such as iron sulphate, were found to have selective herbicidal properties against weeds. Calcium arsenate replaced Paris Green, and by the 1920s arsenic insecticides were widely used on vegetables. The 1930s saw the beginning of the era of synthetic organic compounds, with the introduction of alkyl thiocyanate insecticides, then leading to the discovery in 1939 of the remarkable insecticidal properties of DDT. DDT was followed by the manufacture of other chlorinated hydrocarbons, including aldrin, endrin, hepatchlor and endrin, and the recognition of the herbicidal activity of the phenoxyacetic acids MCPA and 2,4-D. These synthesised products were valued for their persistence in the environment. At the same time, the first organophosphates (OPs), such as parathion and malathion, came into commercial use, some found to be toxic to mammals, but all rapidly degrading in the environment to non-toxic secondary compounds. Later generations of compounds included the carbamates and synthetic pyrethroids, both with low toxicity to humans, modern herbicides and fungicides, and lately neonicotinoids that are chemically similar to nicotine.

Over this time, the use of synthetic pesticides in agriculture has grown steadily, and now amounts to 3.5 billion kg of active ingredient (a.i.) per year. The highest world market growth rates occurred in the 1960s, at 12 per cent per year, later falling back to 2 per cent and below during the 1980s–1990s, then rising to 3 per cent per year by 2014. The value of the global market is now US$45 billion per year (Pretty and Bharucha, 2015). Herbicides account for 42 per cent of sales, insecticides 27 per cent, fungicides 22 per cent, and disinfectants and other agrochemicals 9 per cent. The largest markets are in Europe and Asia (US$12 billion each), Latin America ($10 billion) and North America ($9 billion); the market in the Middle East and Africa is $1.5 billion (all 2012 data). Each synthetic pesticide is costly to manufacture; it has been estimated that the costs to bring a single active ingredient to market are $250 million, companies having synthesised 140,000 compounds to find a single success (Lamberth et al., 2013).

China, USA and Argentina now account for 70 per cent of world pesticide use in agriculture (2.44 billion kg of a.i. annually), with China alone now using half of pesticides worldwide (Table 2.2). Six countries each consume between 50 and 100 M kg (Thailand, Brazil, Italy, France, Canada and Japan) and 13 between 10 and 50 M kg (India, Spain, Germany, Bangladesh, Turkey, South Africa, Russia, Chile, Vietnam, UK, Ghana, Cameroon and Pakistan). In the past 20 years, there have been substantial increases in pesticide use, with consumption in China growing fourfold, Argentina eightfold, Brazil threefold, Bangladesh fivefold and Thailand fourfold. Countries starting from a low base have seen greater increases: Burkina Faso by 50-fold, Ethiopia 13-fold, Ghana 17-fold and Cameroon 8-fold. Aggregate use has been stable in Germany and the USA; there have been notable reductions in use in the UK (down 44 per cent), France (down 38 per cent), Japan (down 32 per cent), Italy (down 26 per cent), Vietnam (down 24 per cent) and Denmark (down 21 per cent). Some of these declines were provoked by policy and regulatory changes, particularly in the European Union.

Table 2.2 also shows some large shifts in the use of insecticides, herbicides and fungicides within the country totals. Several countries in the Organisation for Economic Co-operation and Development (OECD) have seen large declines in the use of insecticides (Denmark, France, Spain, UK, Japan), herbicides (France, Spain, UK) and fungicides (Sweden, Japan). There has been a large increase in herbicide use in Canada (threefold), and substantial reductions in insecticide use in India (down 70 per cent). There have been very large increases in both insecticides and herbicides in Argentina, Thailand, Burkina Faso, Ghana, and increases in all categories in Brazil and Chile. There is no published data on category use in China. The example of Thailand shows the rapid growth in use, rising 9 per cent a year from 1.2 kg/ha to 3.7 kg/ha over 1997 to 2007. The largest amounts are used in the intensive vegetable sector (now averaging 13–20 kg active ingredient for every hectare every year) (Grovermann et al., 2013).

We will show later than at least half of these pesticides are overused or simply not needed to maintain agricultural productivity. In organic agriculture, of course, the assumption is that no inorganic pesticides are necessary. Pesticides are also

TABLE 2.2 Country level agricultural pesticide use (1990 to latest data: 2007–2012)

Country	Latest year (M kg active ingredient) in descending order	Changes in pesticide use over an approximate 20-year period (% change)				Data period (1990s–early 2000s)
		All pesticides	Insecticides	Herbicides	Fungicides	
OECD						
USA	386	101%	88%	95%	43%	1990–07
Italy	63	74%	93%	96%	69%	1990–11
France	62	62%	10%	64%	64%	1990–10
Canada	54	172%	103%	171%	335%	1990–08
Japan	52	68%	74%	102%	54%	2000–11
Spain	40	94%	147%	69%	96%	1990–10
Germany	37	113%	71%	102%	95%	1990–11
UK	16	56%	44%	41%	82%	1990–11
Netherlands	8	99%	53%	80%	88%	1990–10
Denmark	4	79%	15%	112%	37%	1990–11
Sweden	1.8	90%	60%	120%	33%	1990–11
Latin America						
Argentina	265	815%	593%	1,190%	378%	1993–11
Brazil	76	298%	302%	312%	303%	1991–01
Chile	23	263%	349%	228%	201%	1990–11
Asia						
China	1,806	246%	nd	nd	nd	1991–12
Thailand	87	395%	184%	642%	143%	1993–11
India	40	47%	31%	95%	100%	1991–10
Bangladesh	34	489%	2,110%	9,500%	801%	1990–10
Turkey	33	139%	70%	101%	460%	1990–11
Vietnam	19	76%	57%	97%	151%	1994–01
Pakistan	12	129%	148%	42%	51%	1990–01
Sri Lanka	1.3	91%	137%	54%	112%	1991–11
Africa						
South Africa	27	154%	159%	134%	179%	1994–00
Ghana	15	1,683%	591%	5,936%	2,064%	1995–09
Cameroon	11	766%	582%	1,620%	587%	1990–11
Algeria	4	34%	28%	229%	28%	1990–09
Ethiopia	4	1,256%	465%	2,380%	413%	1995–10
Kenya	1.6	44%	27%	64%	47%	1994–01
Burkina Faso	0.8	4,800%	662%	24,800%	nd	1992–11

Sources: Pretty and Bharucha (2015), using China Rural Statistical Report (2013); OECD (2013); FAO Stat (2014a).

Note
nd = no data.

used in other sectors, and use can be substantial: the US Environmental Protection Agency (EPA) (2007) reported total use of 510 M kg of a.i. countrywide, of which 63 per cent was in agriculture, 16 per cent in government activities (for weed control on highways and by forest agencies) and 22 per cent in the home and garden.

In China, pesticide use has increased from 733 M kg in 1990 to 1.806 billion kg in 2012. More than 2.2 billion kg of pesticides are now produced in China, though much is exported. There are 2,000 pesticide companies in China (Zhang *et al.*, 2011). Seven regions each consume more than 100 M kg (Shandong, Hubei, Xinjiang, Hunan, Anhui, Guangdong and Jiangxi), and another ten between 50 and 100 M kg (China Rural Statistical Report, 2013). The grain output per kg varies more than tenfold (from 0.1 to 1.3). Use of pesticide a.i. per hectare also varies considerably by country: it is less than 1 kg/ha in India, Canada and across all Africa, 2–3 kg/ha in the USA and France, and 8–10 kg/ha in the Netherlands, China, New Zealand, Chile and Japan. In the UK, 0.75 kg a.i. were applied per ha in the 1990s; this fell steadily to 0.2 kg in 2012 (Defra, 2014), suggesting farmers had found alternative and more sustainable approaches to pest and disease management.

The use of synthetic pesticides as the frontline strategy against crop pests has long attracted significant concern, but wider use of Integrated Pest Management (IPM) has not yet reduced global pesticide use. There is no clear evidence for example that increased adoption of IPM has led to aggregate changes in pesticide use, with the possible exceptions of Vietnam and Denmark. Even here, patchy data on pesticide use makes it difficult to infer precise long-term trends. Policy changes in Indonesia in the 1980s and adoption of IPM and Farmer Field Schools (FFS) might have changed aggregate pesticide use in the early years, but national level data have not been collected since 1993 (between 1989 and 1993, pesticide use in Indonesia had fallen by 37 per cent). In other contexts, such as in the UK, France, Italy and Japan, it is likely that the economics of farming has been the main driver of reductions in use, with farmers seeking efficiencies by cutting variable costs.

As we shall also see, some changes are accounted for by the large-scale adoption of conservation and no-till agriculture (to some 180 Mha in 2017), and the adoption of genetically modified herbicide-tolerant and insect-resistant crops (also to 150 Mha worldwide), the first allowing for greater use of herbicides, the latter reductions in insecticides. At the aggregate level, this has led to large increases in herbicide use in Argentina and Brazil, but no obvious impact in the USA (Frisvold and Reeves, 2014; though see Benbrook, 2012).

Water and land

Most global agriculture is rainfed, but irrigation accounts for three-quarters of global freshwater use. In coming years, the share of global freshwater available for agriculture is likely to decline as demand grows from industry, power generation and cities. The annual rate of efficiency improvement in agricultural water use between 1990 and 2004 was 1 per cent across both rainfed and irrigated areas. At this rate, the sector will be able to close only 20 per cent of the projected demand–supply gap by 2030. If these gaps between supply and demand become more pronounced, then the negative impacts in countries with high rates of economic growth coupled with high levels of poverty could be significant, in particular India, China, Brazil and South Africa (WRG, 2009). Agricultural run-off from both crop and livestock systems is also a key source of pollution (Conway and Pretty, 1991; Moss, 2008; O'Bannon et al., 2014), thus further reducing the availability of uncontaminated water.

Competition for land, between agricultural and non-agricultural uses and even for different agricultural uses, is increasing (Tilman et al., 2011). In the last decade, a key trend has been the diversion of agricultural land from food to energy crops. In 2013, some 30 per cent of US maize output was diverted into ethanol production (NCGA, 2014), and thus burned in vehicles. In south-east Asia, palm oil production has displaced both food production and natural forest systems (Wicke, 2011; Lee et al., 2014). Many local drivers of conflict relate to control over land and other resources required for livelihoods (Alinovi et al., 2008). These factors lead to the conversion of some non-agricultural land to cultivation, and negative impacts include increased greenhouse gas (GHG) emissions from soils and the removal of carbon sinks (vegetation biomass) and increased fossil fuel use; and increased use of nitrogen fertiliser and the loss of provisioning services to communities who depend on non-agricultural landscapes for food, medicine, fodder, fuel, fibre, cultural identity and spiritual value. Thus, the expansion of agricultural activity into previously uncultivated landscapes has substantial detrimental outcomes. What is needed is to design and manage whole agricultural landscapes better (for both food and environmental services).

Deteriorating soil health poses a grave global challenge in the context of food insecurity, climate change and environmental degradation (McBratney et al., 2014). Soil is a vital asset for agricultural systems; it is also a global sink for carbon. Soil health is diminished through erosion, and by loss of soil carbon, organic matter and nutrients.

Agriculture contributes some 10–12 per cent of global GHG emissions (Tubiello et al., 2013), and in particular between 52 and 84 per cent of global emissions of nitrous oxide and methane (Smith et al., 2008). One estimate of future emissions suggests that by mid-century, methane and nitrous oxide emissions from agriculture could have increased by 50–75 per cent particularly if global diets continue to converge on those typical in the developed North (Popp et al., 2010). Consumption choices are thus part of the problem, particularly in countries transitioning out of historic poverty. Yet, there is much that can be done to improve the efficiency of some forms of animal production, and by diversifying sources of protein within the food system.

Globally, the soil carbon pool is over five times the atmospheric pool and six times the biotic pool (Lal, 2014), and there are 1,500 Pg C in the top one metre of soils globally (Eswaran et al., 1995). A third of global land area is classed as marginal land at high risk of degradation, yet this supports about half of the world population

(Glover and Reganold, 2010). Globally, cultivated soils have lost between 25 and 75 per cent of their organic carbon pool (Lal, 2014); yet some agricultural systems are able to capture and sequester carbon. Agricultural intensification has greatly increased soil nutrient demand for crop production; meeting this demand through synthetic fertilisers is associated with a high energetic, environmental and public health cost (Jones et al., 2013). While nitrogen, phosphorus and potassium fertilisation replenishes some of the nutrients removed by intensive production, many mineral nutrients are inadequately replenished, with negative implications for soil health and nutritional security. The industrial production of fertiliser moves some 120 Mt of atmospheric N to terrestrial and aquatic systems. A further 20 Mt of P are mined annually, and 8–9.5 Mt are released into the world's oceans, some eight times greater than the natural background rate of flux. In certain regions, lack of nutrients in soils remains a key constraint. Many African soils, for example, are nutrient poor, and fertiliser use is low across the continent compared with other regions. The average use of mineral fertilisers does not surpass 6–7 kg of NPK per hectare, against a middle and low income country average of nearly 100 kg ha^{-1}, on land of generally low and declining inherent fertility (Reij and Smaling, 2008). As yields increase, so the net export of nutrients also increases (unless nutrients cycles are closed). Thus, farms in most contexts need to import or fix nutrients to replenish low stocks. As we will show in the following chapters, farmers practising various forms of sustainable intensification have been able to substitute some of these inorganic inputs, including of pesticides, by adding new system elements on their farms, and letting ecosystem services do the work.

Negative food consumption externalities

Agricultural production has grown, yet consumption patterns have changed too, putting pressure back on food and nutritional security. Per capita meat consumption is a good proxy for growing ecological inefficiency in the food system, increasing by 31 per cent per person worldwide since the early 1960s (Figure 2.7). The fastest growth has been in China and Brazil, up 131 per cent and 88 per cent respectively. Such changes also influence global supply chains. Demand in China for animal feedstuffs, for example, places pressures on South American agricultural and non-agricultural ecosystems (Bharucha, 2014). We also know that direct and indirect pressure on agricultural production will come from climate change and depleted ecosystem services fundamental to agricultural success (Costanza et al., 1997, 2014).

By 2050, world population is projected to grow to more than 9 billion, necessitating an increase in total food production by some 70 per cent over early 21st century levels (Godfray et al., 2010; Foresight, 2011). This is why it will not be easy. China's population is predicted to grow by 100 million to 1.46 billion; it will need to increase food supply from domestic sources and/or from imports by 30–50 per cent by 2040–50. For every location, a key challenge centres on policy choices: will past agricultural practices that have brought food production growth continue to succeed, or will new approaches centred on agricultural sustainability be seen as essential?

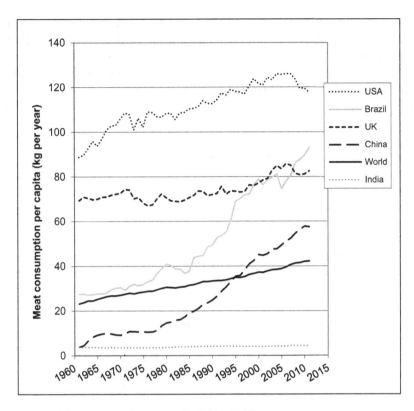

FIGURE 2.7   Meat consumption per capita (1961–2011)

In the modern era, unprecedented increases in food availability in affluent countries, and a host of economic policies, have caused dramatic reductions in the price of food relative to other goods and a falling proportion of household income spent on food. By and large the problems of extreme hunger have been solved. Yet, imbalances remain. In largely solving hunger, affluent economies have now entered mass nutrition transitions. Incidence of obesity has risen in all age cohorts. And overweight and obese body status bring many co-morbidities, in particular cardio-vascular disease and type 2 diabetes. The incidence of diabetes in the UK has increased since 1994 from 2.4 per cent to 6 per cent of adults. Incidence is greatest in the cohort older than 55 years of age, where it has more than doubled to affect one in six of >75 year olds and one in ten of 55–64 year olds. Life expectancy is reduced by ten years by type 2 diabetes. The total health costs of obesity in the UK are rising, and are £20 billion per year (1.5 per cent of GDP of £1.34 trillion) (CMO, 2013; PHE, 2013; Pretty *et al.*, 2015).

The increased access to cheap and energy-dense food in most industrialised countries is a driver of ill-health. In all regions of the world, the proportion of consumed calories as fat has risen, up 32 per cent worldwide since 1961: in

Europe 31 per cent, North America 11 per cent, Asia 71 per cent and Africa only 5 per cent (FAO, 2014a). The UK fast food market is worth £11.4 billion (2012), and comprises 23 per cent of the out-of-home eating sector: sales have increased by 10 per cent since the beginning of the recession in 2008. There are 42,000 fast food and takeaway outlets in England, with an average of 78 per 100,000 people (PHE, 2013). The highest densities (up to 170 per 100,000) are in urban areas, and strong positive associations have recently been shown between an index of multiple deprivation and incidence of fast food outlets (PHE, 2013). There is some evidence from North America that higher rates of obesity occur in communities with high concentrations of fast food outlets (Smoyer-Tomic et al., 2008; Kwate et al., 2009). In the USA, life expectancy has fallen in many southern counties where there is both high income inequality and high levels of obesity (Kulkarni et al., 2011).

A further negative food consumption externality centres on waste: the global cost of food waste has been estimated to be an eye-watering $1 trillion per year (FAO, 2013a). In addition to the actual value of production lost or wasted, environmental costs of wasted production come to some $700 billion, and social costs come to some $900 billion. The annual amount wasted is 1.3 million tonnes. Environmental and social costs include, for example, the cost of emissions, increased water scarcity, soil erosion and risks to biodiversity.

The Sustainable Food Trust (2017) has recently reanalysed the agricultural and food externalities in the UK, and suggest these are much higher than previously thought. For every £1 UK consumers spend on food, additional costs of £1 are incurred and paid by these same consumers. They are paying short term through general and local taxation, drinking water charges, private health care charges, ozone depletion, soil degradation and biodiversity loss. Interestingly taxpayer subsidies to farmers and regulatory and research costs only comprise about 5p in this extra pound (£), more than half of the remainder comprising health costs.

We as consumers pay three times for our food: the price of the food at retail; the cost of environmental damage and clean-up; and the tax revenues spent by governments. These will be hard to change.

Did we say it would be easy?

The impossibility of indefinite growth on a finite planet

Over the past half-century, while world population doubled from 3.5 to more than 7 billion, the size of the world economy grew by nearly fourfold from US$11.2 trillion (at constant 2000 US$) to $42.5 trillion in 2011 (World Bank, 2012). GDP per capita has thus almost doubled. Mean life expectancy has risen from 56 to 70 years, driven strongly by a sharp global fall in under-5 mortality rates from 153 to 51 per 1,000 live births (UNICEF, 2012). Yet these advances in income and health indicators have also brought depletion of natural capital and threats to ecosystem services. Concerns over the finite quantity of natural capital were first articulated by the Club of Rome in *The Limits to Growth* (Meadows et al., 1972), which concluded that the source and sink resources required by the world's economy for continued growth were by definition limited, and that at certain thresholds supply would be outstripped by demand, with negative social and environmental consequences. This would be manifest by rising prices of goods and services together with negative feedback from natural capital and ecosystem services that remained largely unvalued by conventional markets.

The Brundtland Commission (WCED, 1987) then defined sustainable development as forms of economic development that meet human needs and do not damage natural capital in the present or future. The subsequent 1992 Rio conference (and its +10 and +20 events), some international protocols (e.g. Kyoto for carbon dioxide), agreements (e.g. Montreal for ozone), conventions (e.g. Convention on Biodiversity), development targets (e.g. the Millennium and later Sustainable Development Goals), and international and national ecosystem and climate assessments (e.g. by MEA, 2005; IPCC, 2007; NEA, 2011) suggested progress has been made on limiting the impact of human activity on natural capital vital to economies. Yet the iron cage of arithmetic is compelling: per capita consumption of capital continues to rise, as does the total number of consuming people.

The compounded relationship between population (P), affluence (A) (a measure of consumption), technology (T) and their impacts (I) has been variously described by the IPAT and STIRPAT equations (Ehrlich and Ehrlich, 1968; Holdren and Ehrlich, 1974; Dietz and Rosa, 1994; Ehrlich and Ehrlich, 2013; Dietz and O'Neill, 2013). New technologies might offset the multiples of P x A to limit adverse impact, but they have to work very hard (Foresight, 2011; Royal Society, 2012). With both global P and A continuing to rise, the challenge to reduce resulting impacts may be passing beyond human capability. There has been a decline in average energy intensity (by $ spend) by 33 per cent since 1970 – in China down from 8 kg CO_2 per $ of GDP to 3 kg over 1980–2008. Yet, global GDP and population growth have each risen 60–70 per cent over this period (IPCC, 2007). The idea of dematerialisation is alluring, but elusive.

The problem is that gains made by efficiency can be overwhelmed by increases in the size of the economy. This may partly be because of the *rebound effects*, by which efficiency gains free up money which could then be used for further efficiencies or even other non-material consumptive purposes, but which tends to be used to drive more consumption: the Jevons paradox (Jackson, 2017; Sorrell, 2007). In 1970, world CO_2 emissions were 21 Gt CO_{2-eq} yr^{-1}; by 2012 they had risen to 35.6 Gt yr^{-1}. The rate of growth of CO_2 emissions has increased from 0.43 Gt CO_{2-eq} yr^{-1} to 0.92 Gt CO_{2-eq} yr^{-1}. The outcome has been continuing upward pressure on atmospheric CO_2 concentrations: in the pre-industrial era, these were approximately 280 ppm; by 2017 it had reached 400 ppm. At this rate, three decades hence will see the global concentration exceed 450 ppm, an unsafe prospect for humanity. GDP to date has been closely linked to CO_2 production at country level.

GDP closely predicts CO_2 emissions by country (Figure 2.8). Variation between countries suggests a range of efficiencies – those above the predicted relationship

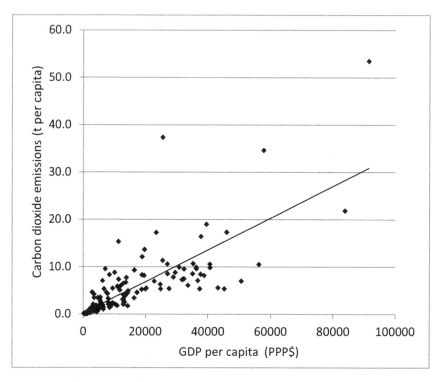

FIGURE 2.8 Relationship between GDP and carbon dioxide emissions at country level (n = 185)

emit more CO_2 per unit of GDP (and thus are carbon inefficient countries), while those below emit less CO_2 than the mean (and are carbon efficient). A slope of zero would indicate complete decoupling of economic performance measured by GDP for carbon emissions. Across all countries, mean per capita CO_2 emissions are 4.4 t CO_2 yr^{-1}. High carbon efficient countries per unit of GDP include Norway, Sweden, Switzerland, Hong Kong, Iceland, France, Singapore, Austria and Israel, and low carbon efficient countries per unit of GDP include Australia, Canada, Kazakhstan, Saudi Arabia, Trinidad and Tobago, UAE, Ukraine and the USA.

A number of global metrics have been developed to demonstrate the impact of human activities on finite Earth. These include the Human Development Index (UNDP), Genuine Progress Indicator (Daly and Cobb, 1989), Ecological Footprints using global hectare equivalents (Moffatt, 2000; WWF, 2010), the Happy Planet Index (NEF, 2012), and planetary boundaries (Rockström *et al.*, 2009). No single approach has captured all source and sink resource use and impact, nor necessarily satisfactorily resolved complexities back to a single metric (such as global earths, hectares or sector boundaries).

However, all conclude that (i) the planet has already exceeded its capacity to supply source and sink resources without resulting in negative feedback loops that reduce the supply of both, and (ii) consumption and population drivers continue to rise, suggesting that the impacts on both environment and economy will also continue to grow. Overshoot has already begun to occur, in which more resources are being used than can be regenerated each year. Yet conventional economic growth is still a primary political goal in most countries. The Royal Society (2012) observed that indefinite growth is impossible in a finite world. Overshoots of consumption could provoke crashes when finite limits are breached.

Tim Jackson (2017) concluded that humans have been *betrayed by affluence*, and Partha Dasgupta (2010) observed that "the rogue word in GDP is gross", as it does not deduct the depreciation of vital natural and social capital assets. The concept of the wealth of nations (Wendell Berry's commonwealth: Berry, 2010) should include measures for natural capital, social capital and individual well-being. It currently does not.

Despite the scientific evidence that anthropogenic sources of carbon are leading to rising atmospheric concentrations that are in turn causing climate change, and that source and sink impacts of consumption are growing, there are a number of denial narratives preventing the necessary changes in policy and behaviour:

i   Economic growth can continue without impact on natural capital and ecosystem services, and such impacts would anyway not result in negative effects on GDP.
ii Increases in GDP linearly improve life satisfaction and well-being through increased consumption.
iii Technological innovation will inevitably produce sufficient changes in the energy intensity of material goods and so will protect source and sink natural capital.
iv Harm caused to natural capital and ecosystem services has no feedback on human well-being.
v   As the poorest countries need to consume more, their pathways to economic development will have to be the same as those adopted by the currently affluent countries.
vi Affluent countries do not need to reduce consumption, as conventional economic growth will eventually deliver environmental benefits.

The macro-political responses have been dismal. The 1992 UN Framework Convention on Climate Change committed signatories to prevent "dangerous anthropogenic interference with the climate system". It had little effect. It had been previously thought that a 2 °C temperature rise would be a "safe guardrail", with such a limit requiring all industrialised countries to cut emissions by 80 per cent on 1990 levels by 2050 (Rogelj *et al.*, 2009). Some conclude that there is virtually no chance of limiting warming to two degrees. It may be that the cuts required are too great, the political will too small, and alternative economic models too disbelieved. Existing pledges by countries are not great enough, and pledges have not led to action. This indicates the scale of the political and technological challenge. Self-interest will clearly slow or prevent transitions, with economic sectors reliant upon fossil fuels likely to continue to promote doubt, following the tobacco industry which has stated that "doubt is our product" (Oreskes and Conway, 2010).

Two problems arise when progress is made: *rebound effects* when money saved by efficiency advances is spent on other damaging goods and services, and *leakage effects* where created natural capital is rapidly spent or lost. The 180 Mha of zero- or low-tillage agriculture adopted in the past 20 years has resulted in the creation of substantial carbon sinks in soils. Yet a return to ploughing would oxidise the sequestered carbon and release it back to the atmosphere.

In the context of agriculture and food production, a central challenge will be to create opportunities and desires for divergence from current pathways and choices. There are three options:

i   Major disruptive and technological innovation followed by widespread and rapid adoption, resulting in the transformation of a large number of economic subsectors.
ii Fixing pro-environmental and low carbon behaviours into societies and economies by cultural moulding.
iii The complete decoupling of the current pursuit of GDP growth as a policy outcome, and increases in investments in the green economy.

It is within this context that agricultural and food systems need to be redesigned. Perhaps, too, this will have positive impacts on other economic sectors, helping to encourage wider transitions towards greener economies. The evidence we outline in the following chapters shows that redesigned agricultural systems can generate multiple, virtuous cycles of social–ecological change. Ecosystems can be improved, people eat better, small businesses thrive, and communities and land knit themselves back together. Agriculture and the environment can flourish together.

3

THE SUSTAINABLE INTENSIFICATION OF AGRICULTURE

At its heart, sustainable intensification is about recognising the fundamental importance of both agricultural and non-agricultural ecosystems, and their links with people. Ecosystem health is a prerequisite for a productive and sustainable agriculture (MEA, 2005; NRC, 2010; Foresight, 2011; NEA, 2011; FAO, 2016a). Agriculture is both driver and recipient of the impacts of global environmental change. What makes agriculture somewhat unique as an economic sector is that it directly affects many of the very assets on which it relies for success. This creates a vital feedback loop from outcomes to inputs. Virtuous or vicious cycles can be created.

Agricultural systems at all levels rely on the value of services flowing from the total stock of assets that they influence and control, and five types of asset – natural, social, human, physical and financial capita – are recognised as being important (Pretty, 2008). There are both advantages and misgivings with the use of the term capital. On the one hand, capital implies an asset, and assets should be cared for, protected and accumulated over long periods. On the other hand, capital can imply easy measurability, substitutability and transferability. If the value of something can be assigned a monetary figure, then it can appear not to matter if it is lost, if a replacement can simply be purchased or sourced from elsewhere. We know, however, that these capitals are not necessarily interchangeable.

Agricultural systems, then, are amended and mostly simplified ecosystems with a variety of important properties that distinguish them from non-cultivated landscapes (Table 3.1). Contemporary, high through-flow agriculture relies for its productivity on simplifying these systems, bringing in external inputs to augment or substitute for natural ecosystem functions, and externalising costs and impacts. Most notably, agriculture in affluent economies relies heavily on energy supplied by fossil fuels directed out of the system (either deliberately for harvests or accidentally through side-effects). Pests are dealt with by the application of synthetic compounds, wastes flow out of the farm in the water supply, or leach into the soil. Where this model can be applied, it can work to increase yields dramatically, but

TABLE 3.1 Properties of natural ecosystems compared with recent agroecosystems typical of affluent economies and sustainable agroecosystems

Property	Natural ecosystem	Recent agroecosystem typical of affluent economies	Sustainable agroecosystem
Productivity	Medium	High	Medium (possibly high)
Species diversity	High	Low	Medium
Output stability	Medium	Low–medium	High
Biomass accumulation	High	Low	Medium–High
Nutrient recycling	Closed	Open	Semi-closed
Trophic relationships	Complex	Simple	Intermediate
Natural population regulation	High	Low	Medium–High
Resilience	High	Low	Medium
Dependence on external inputs	Low	High	Medium
Human displacement of ecological processes	Low	High	Low–Medium
Sustainability	High	Low	High

Source: Gliessman (2005).

usually only in the short term and at increasing cost. By contrast, ecosystem-based sustainable intensification seeks to reintegrate farming into healthy ecosystems, without significantly trading off productivity. The resulting agroecosystems tend to have a positive impact on natural, social and human capital, while unsustainable ones feed back to deplete these assets, leaving fewer for the future. The concept of sustainability does not require that all assets be improved at the same time. One agricultural system that contributes more to these assets than the other can be said to be more sustainable, but there are still likely to be trade-offs with one asset increasing as another falls.

For a transition towards sustainability, renewable sources of energy need to be max-imised, and some energy flows directed towards internal trophic interactions (e.g. to soil organic matter or to non-agricultural biodiversity for arable birds) so as to main-tain other ecosystem functions. These properties suggest a role for agroecological rede-sign of systems so as to produce both food and environmental assets (Hill, 1985, 2014).

As agroecosystems are considerably more simplified than natural ecosystems, some natural properties need to be designed back into systems to decrease losses and improve efficiency. For example, loss of biological diversity (to improve crop and livestock productivity) results in the loss of some ecosystem services, such as pest and disease control. For sustainability, biological diversity needs to be increased to recreate natural control and regulation functions, and to manage pests and diseases rather than seeking to eliminate them. Modern agricultural systems have come to rely on synthetic nutrient inputs obtained from natural sources but requiring high inputs of energy, usually from fossil fuels. These nutrients are often used inefficiently, and result in losses in water and air as nitrate, nitrous oxide or ammonia. To meet principles of sustainability, such nutrient losses need to be reduced to a minimum, recycling and feedback mechanisms introduced and strengthened, and nutrients diverted for capital accumulation (e.g. Thomson *et al.,* 2012). Mature ecosystems are now known to be in a state of dynamic equilibrium, forever moving, shifting and changing. It is this property that buffers against large shocks and stresses. Modern agroecosystems, in prioritising simplification, rigid control and stability, have weak resilience.

In the future, it is likely that many crops will have to be produced under less favourable climatic and economic conditions than those which enabled yield increases during the past century (Glover and Reganold, 2010). In late 2017, the UK government's Natural Capital Committee (2017) sent another warning: "the value of natural capital and the services it provides are often not well-incorporated into decision making processes which rely on market prices. As a result, there is too little investment in natural capital overall, and its wider benefits are not appreciated."

The term sustainable intensification

The desire for agriculture to produce more food without environmental harm, or even positive contributions to natural and social capital, has been reflected in calls for a wide range of different types of more sustainable agriculture: for a doubly green revolution (Conway, 1997), for alternative agriculture (NRC, 1989), for an ever-green revolution (Swaminathan, 2000), for agroecological intensification (Milder *et al.,* 2012; Garbach *et al.,* 2017), for green food systems (Defra, 2012), for greener revolutions (Snapp *et al.,* 2010), for green or evergreen agriculture (Garrity *et al.,* 2010; Koohafkan *et al.,* 2012), for save and grow agriculture (FAO, 2011, 2016c) and diversified agroecological systems (IPES-Food, 2016). Many of these draw on earlier traditions and innovations: permaculture (Mollison, 1988), natural farming and the one-straw revolution (Fukuoka, 1985) and biodynamic agriculture (Koepf, 1989). All centre on the proposition that agricultural and uncultivated systems should no longer be conceived of as separate from each other. In light of the need for the sector to also contribute directly to the resolution of global social–ecological challenges, there have also been calls for nutrition-sensitive (Thompson and Amoroso, 2011), climate-smart (FAO, 2013b) and low carbon (Norse, 2012) agriculture.

Sustainable agricultural systems exhibit a number of key attributes (Pretty, 2008; Royal Society, 2012). They should:

1   Utilise crop varieties and livestock breeds with a high ratio of productivity to use of externally and internally derived inputs.
2   Avoid the unnecessary use of external inputs.
3   Harness agroecological processes such as nutrient cycling, biological nitrogen fixation, allelopathy, predation and parasitism.
4   Minimise use of technologies or practices that have adverse impacts on the environment and human health.

5 Make productive use of human capital in the form of knowledge and capacity to adapt and innovate and social capital to resolve common landscape-scale or system-wide problems (such as water, pest or soil management).

6 Minimise the impacts of system management on externalities such as greenhouse gas emissions, clean water, carbon sequestration, biodiversity, and dispersal of pests, pathogens and weeds.

Agricultural systems emphasising these principles tend to display a number of broad features that distinguish them from the process and outcomes of conventional systems. First, these systems tend to be multifunctional within landscapes and economies (Dobbs and Pretty, 2004; MEA, 2005; IAASTD, 2009). They jointly produce food and other goods for farmers and markets, while contributing to a range of valued public goods, such as clean water, wildlife and habitats, carbon sequestration, flood protection, groundwater recharge, landscape amenity value and leisure and tourism opportunities. In their configuration, they capitalise on the synergies and efficiencies that arise from complex ecosystems, social and economic forces (NRC, 2010). They are negentropic (capital building), the opposite of entropic (Hill, 2015).

Second, these systems are diverse, synergistic and tailored to their particular social–ecological contexts. There are many pathways towards agricultural sustainability, and no single configuration of technologies, inputs and ecological management is more likely to be widely applicable than another. Agricultural sustainability implies the need to fit these factors to the specific circumstances of different agricultural systems (Horlings and Marsden, 2011). Challenges, processes and outcomes will also vary across agricultural sectors: in the UK, for example, Elliot *et al.* (2013) found that livestock and dairy operations transitioning towards sustainability had particular difficulties in reducing pollution while attempting to increase yields.

Third, these systems often involve more complex mixes of domesticated plant and animal species and associated management techniques, requiring greater skills and knowledge by farmers. To increase production efficiently and sustainably, farmers need to understand under what conditions agricultural inputs (such as seed, fertiliser and pesticide) can either complement or contradict biological processes and ecosystem services that inherently support agriculture (Settle and Hama Garba, 2011; Royal Society, 2012). In all cases farmers need to see for themselves that added complexity and increased knowledge inputs can result in substantial net benefits to productivity.

Fourth, these systems depend on new configurations of social capital, comprising relations of trust embodied in social organisations, horizontal and vertical partnerships between institutions, and human capital comprising leadership, ingenuity, management skills and capacity to innovate. Agricultural systems with high levels of social and human assets are able to innovate in the face of uncertainty (Pretty and Ward, 2001; Wennink and Heemskerk, 2004; Hall and Pretty, 2008; Friis-Hansen, 2012) and farmer-to-farmer learning has been shown to be particularly important in implementing the context-specific, knowledge-intensive and regenerative practices of sustainable intensification (Pretty *et al.,* 2011a, 2011b; Settle and Hama Garba, 2011; Rosset and Martínez-Torres, 2012). We shall come back to the critical role of social capital in Chapter 7.

Some conventional thinking about agricultural sustainability has assumed that it implies a net reduction in input use, thus making such systems essentially extensive (requiring more land to produce the same amount of food). Organic systems often accept lower yields per area of land in order to reduce input use and increase their positive impact on natural capital. However, such organic systems may still be efficient if management, knowledge and information are substituted for purchased external inputs. Recent evidence shows that successful agricultural sustainability initiatives and projects arise from shifts in the factors of agricultural production (e.g. from use of fertilisers to nitrogen-fixing legumes; from pesticides to emphasis on natural enemies; from ploughing to zero tillage). A better concept than extensive thus centres on the sustainable intensification of resources, making better use of existing resources (e.g. land, water, biodiversity, knowledge) and technologies (IAASTD, 2009; Royal Society, 2009; NRC, 2010; Foresight, 2011; FAO, 2011; Tilman et al., 2011).

Compatibility of the terms *sustainable* and *intensification* was hinted at in the 1980s (e.g. Raintree and Warner, 1986; Swaminathan, 1989), and then first used in conjunction in a paper examining the status and potential of African agriculture (Pretty, 1997). Until this point, intensification had become synonymous with a type of agriculture that inevitably caused harm whilst producing food (e.g. Collier et al., 1973; Poffenberger and Zurbuchen, 1980; Conway and Barbier, 1990). Equally, *sustainable* was seen as a term to be applied to all that could be good about agriculture. The combination of the terms was an attempt to indicate that desirable ends (more food, better environment) could be achieved by a variety of means. The term was further popularised by its use in a number of key reports: *Reaping the Benefits* (Royal Society, 2009), *The Future of Food and Farming* (Foresight, 2011) and *Save and Grow* (FAO, 2011, 2016c). The term SI was used in the titles of about ten papers a year before 2010, and has since risen to over 100 (Gunton et al., 2016). Sustainable intensification is central to the UN's Sustainable Development Goals for 2030 (Nagothu, 2018).

Many of the initiatives we review in this volume derive from agroecology, the application of ecological principles to the management of both farmed and non-farmed components of agricultural systems. As indicated in the definitions of sustainable intensification, systems that make use of predation, parasitism, allelopathy, herbivory, nitrogen fixation, pollination, trophic dependencies and other ecological processes may be able to develop components that deliver services to the production of crops and livestock. Farming that harnesses ecology by design may thus be able to use fewer or no artificial inputs or compounds (Fukuoka, 1985; Altieri, 1995; Conway, 1997). The term agroecology has also come to be associated with some radical social and political movements, in particular linked to peasant agriculture such as La Via Campesina in Latin America (Rosset et al., 2011; Rosset and Martínez-Torres, 2012; Wibbelmann et al., 2013). A key focus of such agroecosystem management approaches is increased reliance on knowledge and management (or design), complementing and reducing the use of technological inputs.

In 2011, the Food and Agriculture Organization (FAO) of the United Nations also called for a paradigm shift in agriculture, towards sustainable, ecosystem-based production. FAO's latest farming model, Save and Grow, aims to address today's intersecting challenges: raising crop productivity and ensuring food and nutrition security for all, while reducing agriculture's demands on natural resources, its *negative* impacts on the environment and its major contribution to climate change (FAO, 2011, 2016c). The save and grow approach recognises that food security will depend as much on environmental sustainability as it will on raising crop productivity. It seeks to achieve both objectives by using improved varieties, drawing on ecosystem services (such as nutrient cycling, biological nitrogen fixation and pest predation) and minimising the use of farming practices and technologies that degrade the environment, deplete natural resources, add momentum to climate change and harm human health.

Sustainable intensification can be distinguished from former conceptions of agricultural intensification by its explicit emphasis on a wider set of drivers, priorities and goals than solely productivity-enhancement (Table 3.2).

TABLE 3.2 Differences between sustainable intensification and historically conventional forms of agricultural intensification

	Normal or conventional forms of agricultural intensification	*Sustainable intensification*
Primary goals of farmers	Increase crop and livestock yields	Improve yields and incomes, improve natural capital in on- and off-farm landscapes, build knowledge and social capital.
Knowledge development	Tends to be solely expert driven	Collaborations between experts and other stakeholders as key to emergence of agroecological design; participatory research and development leads to new technologies and practices.
Knowledge dissemination	Conventional extension chain from public or private research to farmers	Conventional extension combined with participatory dissemination via peer-to-peer learning.
Stewardship of ecosystem services	Emphasis on provisioning services derived from agricultural landscapes; use of external inputs to substitute for regulating and supporting services; interactions with surrounding non-agricultural landscapes treated as externalities	Greater appreciation of the contribution of multiple ecosystem services provided by agricultural landscapes and awareness of the two-way relationship between agricultural and non-agricultural components of landscapes.

Source: Pretty and Bharucha (2014).

Differing views on sustainable intensification

All terminology brings controversy. Garnett and Godfray (2012) reviewed key contentions and debates surrounding sustainable intensification, classifying these into three groups. The first relates to the *vision and mode* of SI, wherein the term is assumed to prescribe particular forms of agriculture deemed unsuitable for various reasons. The second questions the *rationale* for SI, and a third set of questions relates to the *conceptual basis* of SI: which is more important, *sustainable* or *intensification* and how do they relate to one another?

One contention relates to the potential for SI to be interpreted simply as a productivist project. Much criticism of conventional (recent, normal: see Kuhn, 1970) agriculture centres on concerns over large-scale industrial monocultures concerned only with increasing yields and the gross productivity of systems. But a good agriculture would also be efficient in its use of resources, and equitable in providing access to its food produced (Foresight, 2011). Some have said SI is simply a Trojan Horse designed to smuggle more biotech and genetic modification into agriculture. Others have asked whether the concept represents a sufficiently radical departure from 'business-as-usual'. Some have highlighted distinct and competing strong and weak interpretations of sustainable intensification, or the need to establish one agreed set of indicators (Marsden, 2014; Mahon *et al.*, 2017). Weak interpretations may be open to the charge of promoting an apparent oxymoron (Lang and Barling, 2012) that may simply be used as a greenwash. Implicit in some of these concerns is the notion of an association between large-scale and particular technologies, and a distinction between the values of large and small farms, with an implicit preference for the latter. These point to a tension between different conceptions of what is good in agriculture, and reveal some of the complexity that sustainable intensification must navigate. We will show in Chapters 5 and 6 that SI can work in both small and large farms, and in developing as well as industrialised contexts.

In practice, it may not be easy to distinguish between approaches. For example, Conservation Agriculture (CA) and Integrated Pest Management (IPM) can both be thought of variously as sustainable intensification, also as agroecology, as climate-smart agriculture, as ecological intensification or simply as a greener agriculture (Gliessman, 2014; Kassam *et al.*, 2009). These terms reflect differing priorities on agricultural inputs and outputs but "all will have to engage with the reality that there are hard trade-offs between different desirable outcomes and uncomfortable choices for all stakeholders" (Garnett and Godfray, 2012).

Going beyond privileging any particular agricultural technology, and by focusing only on desirable social–ecological outcomes, there is a need to evaluate all technologies, approaches or practices pragmatically and empirically. We should judge each on its merits: does it produce more food per unit of resource; and does it do so without harm to the environment? It remains clear, though, that better agricultural and food systems could be imagined by reducing food waste, increasing community engagement and reducing inequity, regardless of the forms of production in fields and farms, or indeed the terminology. As important in agricultural systems to farmers and workers are returns to labour, and the distribution of benefits between women and men.

But even the openness of SI throws some difficult questions into relief. Defining sustainability is hard. As with different versions of sustainability, it is possible to argue that SI has light and dark green interpretations. Defining boundaries – between agriculture and other economic sectors or around units of landscape (farms, watersheds, landscapes) or around time spans (five-year plans, decades, across generations) – is also difficult because of incomplete knowledge, continually evolving conditions and diverse human values. Again, outcomes are important: social and political transformations may be needed to ensure that yield increases delivered through SI do actually reduce hunger and poverty (Holt-Giménez and Altieri, 2013).

Terminology can hide variations in practice, and often sustainability outcomes. For example, IPM constitutes a wide range of methods, practices and technologies available to reduce pest–weed–disease threats. Some approaches centre on agroecological management and habitat design, using the services of biodiversity on- and off-farm. Others simply centre on the scheduling of applied pesticides. The US National Research Council (2010) noted that for some farmers in the USA, "IPM means simply scheduling pesticide applications based on monitoring and established economic thresholds; others use more integrated IPM ... with pesticide use as a last resort."

There may also be ambiguity about *what* is being intensified. Jacobsen *et al.* (2013) argued, "Many arguments about feeding the world assume that we need more of our current, western diet, but it should be obvious that the world's population can better be fed, both agriculturally, environmentally and with respect to human health, with a diet different from what is most common in the developed world today." It is also not always accepted that yields need to be increased (Tomlinson, 2013). Elliot *et al.* (2013) point out that in certain cases, SI "may not be an appropriate strategy [because] other ecosystem functions may be valued more highly than increases in food production (e.g. water quality, carbon storage, landscape quality)".

Another common objection made about many agroecological approaches for sustainable intensification is their perceived need for increased labour (Tripp *et al.*, 2005). However, sustainability concerns are highly site-specific: in some cases more labour is not needed; in others higher labour requirements are seen as valuable contributions to local economies (De Schutter and Vanloqueren, 2011). In some contexts, farm labour is highly limiting, especially where HIV-AIDS has removed a large proportion of the active population; in other contexts, there is plentiful labour available as there are few other employment opportunities in the economy. Successful systems of sustainable intensification by definition fit solutions to local needs and contexts, and so thus take account of labour availability. In Kenya and Tanzania, for example, female owners of raised beds for vegetable production employ local people to work on vegetable cultivation and marketing (Muhanji *et al.*, 2011). Labour for crop and livestock management is thus not necessarily a constraint on new technologies. Again, it depends on local context.

In Burkina Faso, work groups of young men have emerged to aid soil conservation. *Tassa* and *zai* planting pits are best suited to landholdings where family labour is available, or where farm hands can be hired (Reij and Smaling, 2008; Sawadogo, 2011). The technique has led to a network of young day labourers who have mastered this technique. Owing to the success of land rehabilitation, farmers are increasingly buying degraded land for improvement, and paying labourers to dig *zai* pits and construct the rock walls and half-moon structures, which have transformed productivity. This is one of the reasons why more than 3 Mha of land are now rehabilitated and productive. In other contexts, though, shifts to sustainable systems, such as incorporating agroforestry into maize systems in Africa, has led both to reduced and increased labour requirements, depending on the local social and ecological context.

Addressing declining efficiencies: the case of China

Agricultural change in China is instructive: there have been goods and bads. Improvements in agricultural output in China have been largely driven by increases in consumption of four factors of production: fertilisers, pesticides, fuel and water (Table 3.3). At the same time, the efficiency of these resources has declined: only the water-use efficiency for grain production has increased (Figure 3.1). Some of the required increases in food production could come from further increases in use efficiency of these factors of production, but with a declining marginal efficiency, costs to farmers and the wider economy will rise (Norse and Ju, 2015). Concurrently, Chinese arable land area is slightly changed and limited in supply: over 2009–2013 the arable land area has remained at about 135 Mha (MOLR, 2010–2014). It would appear that further increases in land area for grain cropping are not a viable option. Further development would move into less favourable arable land and cause a loss of ecosystem services from the conversion of valuable habitats and natural systems. At the same time, some arable land is being lost to urban encroachment (Su *et al.*, 2011).

A further cost arises from direct negative impacts of some agricultural practices on natural capital and human health. As we showed in Chapter 2, China is now the largest consumer of pesticides worldwide, more than tripling use since 1990 to 1.81 billion kg of the current world annual total of 3.5 billion kg (Zhang *et al.*, 2011). We shall see later that evidence from integrated pest management projects in Asia and Africa shows that at least 50 per cent of pesticides applied (by weight) are overused, resulting in each kg of active ingredient imposing US$4–19 of

TABLE 3.3 Changes in four factors of production, China

	Current annual use	*Increase over period*	*Years*
Fertilisers	58.4 Mt	+227%	1990–2012
Pesticides	1.806 billion kg	+236%	1990–2012
Fuel	36.5 Mt	+64%	2000–2013
Water	392 Mt	+4%	2000–2013

Source: Li *et al.* (2016), using NBS (2013, 2014); MOA (2000–2014).

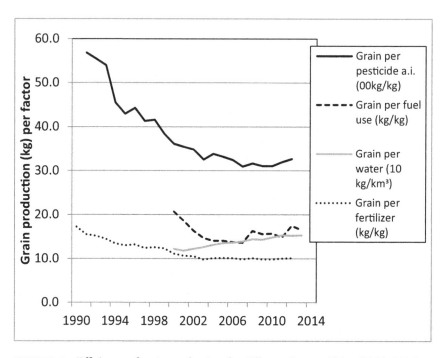

FIGURE 3.1 Efficiency of grain production for different factors, China (1990–2013)

Source: Li *et al.* (2016), using National Bureau of Statistics (NBS) of China (2013, 2014); Ministry of Agriculture (MOA) of the People's Republic of China (2000–2014).

Notes

Pesticides include all herbicides, insecticides and fungicides; grains include all major and minor grains.

external costs to the environment and human health (Norse *et al.*, 2001; Pretty and Bharucha, 2015).

Fertiliser use in China has increased by 227 per cent to 58Mt annually, yet not all applied nutrients are taken up by plants or microbial communities and thus escape and contribute to the costly eutrophication of aquatic ecosystems (Norse *et al.*, 2001) and nitrous oxide emissions (Yu *et al.*, 2015). Soil degradation affects agriculture's ability to continue to be productive as well as adversely impacting ecosystems. In China, a total of 295 Mha of land is prone to accelerated soil erosion (MOEP, 2014), and in some principal rivers annual soil sediment losses can exceed 300Mt (Table 3.4). For arable lands, topsoil rich in organic matter and nutrients is usually the first to be eroded, thus resulting in considerable flux of carbon to the atmosphere and negative impacts on soil productivity (Wang *et al.*, 2006).

The challenge is substantial: China must produce 30–50 per cent more food during the next 25–35 years, yet it must also find ways to adopt agricultural innovations that improve natural capital and ecosystem services. This is an

TABLE 3.4 Annual soil sediment losses from the principal rivers in China

River	Annual average	
	(1950–1995)	*2013*
	Total amount of erosion (Mt)	*Total amount of erosion (Mt)*
Yangtze River	23,870	550
Yellow River	16,000	830
Zhujiang River	2,200	670
Songhuajiang River	190	320
Liaohe River	1,530	340
Talimu River	1,300	1,040

Sources: Li *et al.* (2016); MOWR (2013).

example of where sustainable intensification should play a vital role in increasing yields without adverse environmental impact *and* without the cultivation of more land.

SI can be distinguished from earlier conceptions of agricultural intensification as a result of its explicit emphasis on wider environmental and health outcomes. It is often achieved from shifts in the factors of agricultural production (e.g. from use of fertilisers to nitrogen-fixing legumes; from pesticides to emphasis on natural enemies and on creating disease-suppressive soils with favourable micro-biota; from ploughing to no-till or NT) rather than simply intensifying the use of existing inputs. A number of approaches have been identified for the SI of agroecosystems, including crop variety improvements, IPM, management intensive rotational grazing systems, integrated nutrient management, conservation agriculture and agroforestry systems (Ellis and Wang, 1997; Smith, 2013; Barzman *et al.*, 2014; Pretty and Bharucha, 2014, 2015).

The real costs of pesticide use

Pesticides are intended to be hazardous, and their real costs illustrate the often hidden harm of non-sustainable agricultural systems. The value of pesticides lies in their ability to kill unwanted organisms. Most act by interfering with biochemical and physiological processes that are common to a wide range of organisms, including not only pests, weeds and fungi, but also wildlife and humans. The risks differ from compound to compound, and much of the information on their side-effects remains widely contested.

Most economic studies assessing the benefits of pesticides are based on a comparison of two scenarios: current use versus complete zero. These ask how much reduced pesticide use would cost farmers and the agricultural industry (e.g. Knutson *et al.*, 1998; Schmitz, 2001), and have concluded that substantial costs would arise from yield reductions, the need for additional food imports and greater consumer risks through exposure to mycotoxins on foods. For the USA, the costs following a complete ban of pesticides were put at $18 billion (Knutson *et al.*, 1998), and 1 per cent of GDP in Germany after a 75 per cent ban (Schmitz, 2001). Objections subsequently raised to these macro studies centred on the lack of data to describe the relationships between pest/disease ecosystems and the economic system, the reliance on expert opinion, data only derived from research stations and the failure to account for impacts on human health and the environment (negative externalities) (Pearce and Tinch, 1998; Pretty and Waibel, 2005; Leach and Mumford, 2011).

Pesticide externalities show features commonly found across the agricultural sector: (i) their costs are often neglected; (ii) they often occur with a time lag; (iii) they often damage groups whose interests are not well represented; and (iv) the identity of the producer of the externality is not always known. Pesticide externalities thus result in suboptimal economic and policy outcomes.

Country level externalities have been reported for China, Germany, the UK and USA for five categories (Table 3.5), though these are almost certainly underestimates owing to differing risks per product, poor understanding of chronic effects (e.g. in cancer causation or population level effects in ecosystems), weak monitoring systems and misdiagnoses by doctors of health effects. This framework includes only externalities, those costs passed on to the rest of society through the actions

TABLE 3.5 Cost Category Framework For Assessing Full Costs Of Pesticide Use (Million US$ Per Year, 2000)

Damage costs	China[1]	Germany	UK	USA
1 Drinking water treatment costs	nd	104	215	1,059
2 Health costs to humans (farmers, farm workers, rural residents, food consumers)	500–1300	17	2[2]	157
3 Pollution incidents in watercourses, fish deaths, monitoring costs and revenue losses in aquaculture and fishing industries	nd	60	7	153
4 Negative effects on on- and off-farm biodiversity (fish, beneficials, wildlife, bees, domestic pets)	200–500	10	75	331
5 Negative effects on climate from energy costs of manufacture of pesticides	148	4	3	55
TOTALS	848–1,948	195	302	1,755

Notes
1 China costs are just for rice cultivation.
2 Does not include any costs of chronic health problems.
nd = no data.

of agriculture. Koleva and Schneider (2009) later put the external costs of pesticide use in the USA as much higher (including home and garden use): $12.5 billion annually ($9.5 billion for human health, $3 billion for the environment), and amounting to $42 of costs imposed from every hectare of agricultural land.

Additional private costs borne by farmers themselves are not included, such as from increased pest or weed resistance from the overuse of pesticides, or for training in the use, storage and disposal of pesticides. There also remain unmeasured distributional problems: for example, insect outbreaks arising from pesticide overuse can affect all farmers, even those not using pesticides. There may also be trade-offs: minimum tillage in Asian rice has, for example, resulted in large water savings, reductions in labour needed for land preparation, some yield increases, yet also increases in herbicide use (Kartaatmadja et al., 2004).

The continued use of poorly regulated generics, particularly in developing countries, has led to the 'lock-in' of several obsolete pesticides (Popp et al., 2013). In Africa, it is estimated that 10 per cent of products used are in World Health Organisation Class 1a (extremely hazardous) and 1b (highly hazardous). These are mostly banned in industrialised countries. Protective clothing is almost unknown, and poisoning incidence and hospitalisation is common in some regions (Williamson et al., 2008). Jepson et al. (2014) surveyed 1,700 individuals growing 22 crops and using 31 pesticide compounds, and recorded symptoms of cholinesterase inhibition, developmental toxicity, impairment of thyroid function and depressed red blood cell counts. The externalities of pesticides in Mali are estimated to be 40 per cent of the costs paid by farmers for pesticide products.

Leach and Mumford (2008, 2011) also calculated the costs per kg of active ingredient for the UK, USA and Germany. They show that external costs range from €3–15 ($4–19) per kg active ingredient. Earlier research had shown costs for rice cultivation in China to be $3–6.5 per kg (€2.4–5) (Praneetvatakul and Waibel, 2006; Leach and Mumford, 2011). For Thailand, Praneetvatakul et al. (2013) calculated the negative externalities of rice to be $19/ha, but rising to $106/ha for intensive vegetables. The active ingredient costs are thus $7.3/kg for rice, and $8.2/kg for vegetables. These costs put crude pesticide externalities worldwide in the range of $10–60 billion (for use of 3.5 billion kg and for a market size of $45 billion). This indicates that the benefits of pesticide use for pest, disease and weed control and their ease of use should be set in the context of the costs of unintended side-effects. Such an understanding can thus frame the potential benefits of adopting alternative methods and practices of IPM that produce pest control with less pesticide use.

A recent review of pesticides and impacts on health confirms there are no satisfactory data on pesticide impacts worldwide (Andersson et al., 2014). A comprehensive analysis of 122 studies published post-2000 showed specific cohorts, especially applicators and farmers, often had significantly higher odds ratios for a range of cancers, risks of dementias and respiratory symptoms. Athukorala et al. (2010) reported that 4–7 per cent of agricultural workers suffer ill-health from pesticides each year in Sri Lanka, Costa Rica and Nicaragua.

TABLE 3.6 Benefits And Health Costs Of Three Pest Management Strategies In Irrigated Rice, Philippines

Pest management strategy	Agricultural returns, excluding health costs (Pesos/ha)	Health costs (Pesos/ha)	Net benefit (Pesos/ha)
Complete protection: standard practice of 9 pesticide sprays per season	11,850	7,500	4350
Economic threshold: treatment only when threshold passed, usually no more than two applications used	12,800	1,190	11,610
IPM: pest control emphasises predator preservation and habitat management, alternative hosts and resistant varieties	14,000	0	14,000

Source: Pingali and Roger (1995).

When asked, farmers often say they are willing to pay to protect against pesticide-induced ill-health, but then generally do not. Pingali and Roger (1995) calculated the human health costs of pesticide use in irrigated rice systems of the Philippines, and compared the economics of three pest control strategies: complete protection comprising nine sprays per season, economic threshold decisions involving two sprays per season, and IPM with no pesticides. The complete protection cost most in terms of ill-health and returned the least agricultural returns per hectare (Table 3.6). Pingali and Roger (1995) concluded: "the value of crops lost to pests is invariably lower than the cost of treating pesticide-related illness and the associated loss in farmer productivity. When health costs are factored in, the natural control option is the most profitable pest management strategy". Finding effective IPM approaches that reduce pesticide use would thus have important effects on rural public health. This again illustrates the potential for good from implementation of more sustainable approaches to agriculture.

Now we turn in the next chapters to the evidence for the successes in the development, use and spread of sustainable intensification.

4

DOES SUSTAINABLE INTENSIFICATION WORK?

We promised at the start of this book that we would marshal evidence for sustainable intensification. We begin in this chapter with meta-level analyses where multiple farms, projects and initiatives of different types have been analysed for their impact. There are three key questions. Can the sustainable intensification (SI) of agricultural systems work? Can SI, at the first and production stage of food chains, produce more food, fibre and other valued products whilst improving natural capital? Is it possible to produce more whilst not trading off harm to key renewable capital assets?

Some caveats: documenting and evaluating evidence of sustainable intensification is complex, and sometimes contentious. We have already summarised some of the debates about sustainable intensification as a concept. These aside, there are real difficulties with operationalising the concept of sustainability and evaluating sustainable agroecosystems.

First, as we have outlined earlier, sustainable intensification is an umbrella concept, with a diverse array of initiatives, processes and technologies involved. The inclusivity of the approach means that it is difficult to bound evaluations. It is rarely an easy matter to decide what to include and exclude in any evaluation of a sustainable intensification initiative. This is partly because agroecological approaches involve multiple practices, adapted from place to place depending on farmer and community needs. There may be no clear conceptual, methodological or practical dividing line between a new or alternative practice and a conventional or normal practice. Outcomes are key; pathways differ. Agroecological approaches are also often multiple, consisting of packages of technologies that can be applied to varying extents across different types of farm. For example, depending on need and ability, farmers may apply agroecological principles to industrial farms, or introduce mechanisation and inorganic inputs into otherwise agroecologically managed farms (Milder et al., 2012; Lampkin et al., 2015). In sum then, identifying the boundary conditions for sustainable intensification is a challenge. Thoughtful work on developing new systems of metrics and evaluation frameworks may help (Smith *et al.*, 2017), but it is difficult, and not necessarily desirable, to measure everything quantitatively. Relations of trust, social connections, human capital and innovation can all be proxied, but only imperfectly. Finally, there is not yet sufficient data on how different sustainable intensification strategies (e.g. using agroecological methods) might meet aggregate regional and global goals for food security and economic success. What is the potential for scaling up, and what might the challenges be?

Each of these features means that defining the exact scope of existing sustainable intensification is difficult. Though we may be able to measure how many hectares are covered by particular technologies, or how many farmers adopt particular practices, the sum of the global transition to agroecological approaches is difficult to judge. In 2012, Milder *et al.* estimated that globally some 200 Mha of agricultural land were being cultivated under some form of agroecological regime. At the same time, smallholder production is particularly dependent on healthy ecosystems on and around farms (IFAD and UNEP, 2013), and it has been estimated that half the world's smallholders could be practising some forms of sustainable intensification of agriculture (Altieri and Toledo, 2011; IFAD and UNEP, 2013).

Most evaluations of sustainable intensification compare improved production practices with either a baseline or conventional good practice. Studies seek to demonstrate simultaneous improvements to yields and environmental outcomes, but results can be sensitive to the variables and parameters selected to capture environmental improvements, the timescales involved and any weightings used (Elliot *et al.*, 2013).

Methods matter too. Some assessments have been found to suffer from methodological flaws (see Milder *et al.*, 2012). First, despite the heterogeneity of practices involved in any intensification strategy, assessments often focus on yields from specific, labelled approaches – such as Conservation Agriculture, agroforestry or Integrated Pest Management. Analysis of distinct approaches is also difficult. For example, evidence on outcomes from both Conservation Agriculture and the Systems of Crop and Rice Intensification is mixed, and debate on the general applicability and scalability of these approaches has been "high profile, sustained and at times acrimonious and emotive" (Sumberg *et al.*, 2013; see also Glover, 2011). Second, syntheses, meta-analyses and overviews have so far focused primarily on yield increases rather than on multiple outcomes and benefits (but see Pretty *et al.*, 2006, 2011a, 2011b; Milder *et al.*, 2012). Given that sustainable intensification is explicitly about much more than increased yields, much remains eclipsed by this approach. New approaches are often needed to explore all benefits, trade-offs and synergies that come with a new package of technologies. Studying packages and their impacts in their entirety calls for a fundamental departure from conventional research, focusing less on single outcomes (e.g. yield) or single media (e.g. outcomes for water, or air, or C or N in isolation) (Rosenstock *et al.*, 2014). A related difficulty is that research on sustainability of ecosystems is itself in need of further development. Biodiversity is an example. There are a variety of measures on agricultural biodiversity, but data and metrics are spread across various disciplines. This comes in the way of developing a clear understanding of trends (Bioversity International, 2016).

Finally, there is also the question of what has been entirely eclipsed or ignored. As we outline in Chapter 5, small patches and home gardens are an important source of food security and ecosystem services across the world, yet there is little effort to target these as an important frontier in agricultural research and development. Existing research has focused more on tropical home gardens relative to temperate ones (Galhena *et al.*, 2013; but see Vogl and Vogl-Lukasser (2003) and Calvet-Mir *et al.* (2012) for studies from Austria and Spain). Agricultural extension and research have also largely ignored patches that fall outside regular field boundaries (Pretty *et al.*, 2011b), despite calls that more attention be given to home gardens, kitchen gardens and other small-scale, subsistence-oriented enterprises in agricultural research, development and extension.

Having noted these limits in the existing base of evidence, it is important to highlight that existing studies which do focus on yields show significant positive impacts of a variety of sustainable intensification packages (see, for example, some of the evidence we highlight in Chapters 5 and 6). Yields, though a crude measure, are a reason for optimism. Studies of system redesign in particular have shown the beneficial outcomes of both–and approaches rather than either–or. In other words, approaches that seek to build social and natural capital while increasing yields tend to work better than approaches skewed simply towards increased yields or particular ecosystem components.

Analyses tend to be of two types: involving temporal measures of impact (changes over time at the same location, sometimes called longitudinal) and spatial measures (changes measured at the same point but with different treatments, sometimes called latitudinal).

Farmers adopting various sustainable intensification approaches have been able to increase food outputs by sustainable intensification in two ways. The first is *multiplicative* – by which yields per hectare have increased by combining use of new and improved varieties with changes to agronomic–agroecological management. The second is improved food outputs by *additive* means – by which diversification of farms resulted in the emergence of a range of new crops, livestock or fish that added to the existing staples or vegetables already being cultivated. These additive system components may include:

i   Aquaculture for fish raising (in fish ponds or concrete tanks).
ii   Small patches of land used for raised beds and vegetable cultivation.
iii   Rehabilitation of formerly degraded land.
iv   Fodder grasses and shrubs that provide food for livestock (and increase milk productivity).
v   Raising of chickens, and zero-grazed sheep and goats.
vi   New crops or trees brought into rotations with staple (e.g. maize, sorghum) yields not affected, such as pigeonpea, soyabean, indigenous trees.
vii   Adoption of short-maturing varieties (e.g. sweet potato, cassava) that permit the cultivation of two crops per year instead of one.

Impacts on productivity

One of the earliest large-scale assessments was commissioned in 1989 by the US National Research Council (NRC). This resulted in the seminal *Alternative Agriculture*. Partly driven by increased costs of fertiliser and pesticide inputs, plus growing scarcity of natural resources (such as groundwater for irrigation) and continued soil erosion, farmers had been adopting novel approaches in a wide variety of farm systems. The NRC noted that Alternative Agriculture was "not a single system of farming practices", that they were compatible with large and small farms, and were often diversified. Such alternative agricultural systems used crop rotations, IPM, soil and water conserving tillage, animal production systems that emphasised disease prevention without antibiotics, and genetic improvement of crops to resist pests and disease and use nutrients more efficiently. Well-measured alternative farming systems nearly always used less synthetic chemical pesticide, fertiliser and antibiotic per unit of production than comparable conventional farms (NRC, 1989). They also required more information, trained labour and management skills per unit of production.

The NRC (1989) commissioned 11 detailed case studies of 14 farms as exemplars of effective and different approaches to achieving similar aims: economically successful farms with a positive impact on natural capital. The NRC (2010) later conducted follow-up studies in 2008 on ten of the original farms. These included integrated crop–livestock enterprises, fruit and vegetable farms, one beef cattle ranch and one rice farm. After 22 years, there were six common features of these farms:

1   All farms emphasised the importance of maintaining and building up their natural resource base and maximising the use of internal resources.
2   All farmers emphasised the values of environmental sustainability and the importance of closed nutrient cycles.
3   The crop farms engaged in careful soil management, the use of crop rotations and cover crops; the livestock farms continued with management practices that did not use hormones or antibiotics.
4   More farmers participated in non-traditional commodity and direct sales markets (via farmers markets and/or the Internet); some sold at a premium with labelled traits and products (e.g. organic, naturally raised livestock).
5   Most farms relied heavily on family members for labour and management.
6   The challenges and threats centred on rising land and rental values associated with urban development pressure, the availability of water and the spread of new weed species.

It is in developing countries that some of the most significant progress towards sustainable intensification has been made in the past two decades (Pretty *et al.*, 2006, 2011a, 2011b; Pretty and Bharucha, 2014, 2015). The largest study to date comprised the analysis of 286 projects in 57 countries. In all, some 12.6 million farmers on 37 million hectares were engaged in redesign transitions involving sustainable intensification. At the same time, this comprised about 3 per cent of the total cultivated area (1.14 Mha) in developing countries. In 68 randomly re-sampled projects from the original study, there was a 54 per cent increase over the subsequent four years in the number of farmers, and 45 per cent increase in the number of hectares (Pretty, 2008). These re-surveyed projects comprised 60 per cent of the farmers and 44 per cent of the hectares in the original sample of projects.

For the 360 reliable yield comparisons from 198 of the projects, the mean relative yield increase was 79 per cent across the very wide variety of systems and crop types. However, there was a wide spread in results. While 25 per cent of projects reported relative yields of more than 2.0 (i.e. 100 per cent increase), half of all the projects had yield increases of between 18 per cent and 100 per cent. Though geometric mean is a better indicator of the average for data with a positive skew, this still shows a 64 per cent increase in yield for eight different crop groups (Figure 4.1).

Table 4.1 summarises the changes in yields, along with numbers of farmers and hectares, for eight different smallholder agroecosytems: irrigated, wetland rice, rainfed humid, rainfed highland, rainfed dry/cold, dualistic mixed, coastal artisanal and urban-based. The mean farm size for these 12.6 million famers was 2.9 hectares.

As we have seen in earlier chapters, progress towards agricultural intensification has been uneven globally. The green revolutions of the late 20th century brought significant change to aggregate yields of important staples in Asia and South America. Yields across Africa have not kept pace. Yet, here too there have been important developments over the past few decades. The UK Government Office

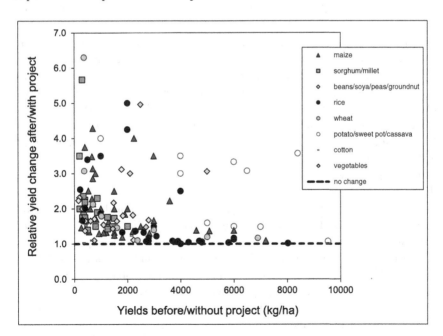

FIGURE 4.1   Relative changes in crop yields under sustainable intensification, 198 projects, 57 countries (Pretty *et al.*, 2006)

TABLE 4.1 Summary of adoption and impact of agricultural sustainability technologies and practices on 286 projects in 57 countries (early 2000s)

FAO smallholder farm system category[1]	Number of farmers adopting	Number of hectares under sustainable intensification	Average % increase in crop yields[2]
1 Irrigated	177,287	357,940	129.8 (±21.5)
2 Wetland rice	8,711,236	7,007,564	22.3 (±2.8)
3 Rainfed humid	1,704,958	1,081,071	102.2 (±9.0)
4 Rainfed highland	401,699	725,535	107.3 (±14.7)
5 Rainfed dry/cold	604,804	737,896	99.2 (±12.5)
6 Dualistic mixed	537,311	26,846,750	76.5 (±12.6)
7 Coastal artisanal	220,000	160,000	62.0 (±20.0)
8 Urban-based and kitchen garden	207,479	36,147	146.0 (±32.9)
All projects	12,564,774	36,952,903	79.2 (±4.5)

Notes

1 Smallholder farm categories from Dixon and Gulliver (2001).

2 Yield data from 360 crop-project combinations; reported as % increase (thus a 100% increase is a doubling of yields). Standard errors are in brackets.

of Science Foresight programme commissioned reviews and analyses from 40 projects in 20 countries of Africa where sustainable intensification had been developed or practised in the 2000s (Pretty *et al.*, 2011a; Pretty *et al.*, 2014). The cases comprised crop improvements, agroforestry and soil conservation, conservation agriculture, integrated pest management, horticultural intensification, livestock and fodder crops integration, aquaculture, and novel policies and partnerships. By early 2010, these projects had recorded benefits for 10.4 million farmers and their families on approximately 12.75 million hectares. Across the projects, yields of crops rose on average by a factor of 2.13 (i.e. slightly more than doubled) (Table 4.2). The timescale for these improvements varied from three to ten years. It was estimated that this resulted in an increase in aggregate food production of 5.79 million tonnes per year, equivalent to 557 kg net per farming household (in all the projects). We shall learn of how some of these changes were achieved in the next chapter.

In France, the Institut pour Agriculture Durable (2011) called for a new European agriculture based around maintaining healthy soil, biodiversity, appropriate fertilisation and appropriate plant protection techniques. Deploying these "helps protect the environment whilst producing more, better and in another way". Testing 26 indicators classed into seven themes (economic viability, social viability, input efficiency, soil quality, water quality, GHG emissions and biodiversity) across 160 different types of farm, the IAD team found that positive ecological externalities can both be achieved and measured. Together, these indicators comprise a comprehensive scorecard that can be applied to test progress towards the production

TABLE 4.2 Summary of productivity outcomes from SI projects in Africa

Thematic focus	Area improved (ha)	Mean yield increased (ratio)	Net multiplicative increase in food production (1,000 tonnes/year)
Crop variety and system improvements	391,060	2.18	292
Agroforestry and soil conservation	3,385,000	1.96	747
Conservation agriculture	26,057	2.20	11
Integrated pest management	3,327,000	2.24	1,418
Horticulture and small-scale agriculture	510	nd	nd
Livestock and fodder crops	303,025	nd	nd
Novel regional and national partnerships and policies	5,319,840	2.05	3,318
Aquaculture	523	nd	nd
Total	12,753,000	2.13	5,786

Sources: Foresight (2011); Pretty *et al.* (2014).

Notes
nd = no data, largely because horticulture, livestock and aquaculture are additive components to systems, increasing total food production but not necessarily yields.

of positive ecological externalities as well as maintenance of productivity. In the UK, Elliot *et al.* (2013) explored outcomes across 20 farms, of which four appeared to have achieved yield increases alongside environmental improvements (denoted by reduced pollution and increased biodiversity). The study showed some of the first evidence of SI in the UK, achieved through a mixture of new technologies (improved genetics and precision farming), new practices (zero tillage (ZT) and improved water management), diversification (the installation of small-scale energy generation) and the application of available agri-environmental schemes.

Improving environmental externalities

Environmental externalities have often been shown to be positive as a result of sustainable intensification. Carbon content of soils is improved where legumes and shrubs are used, and where conservation agriculture increases the return of organic residues to the soil. Legumes also fix nitrogen in soils, thereby reducing the need for inorganic fertiliser on subsequent crops. In IPM-based projects, most have seen reductions in synthetic pesticide use (e.g. in cotton and vegetables in Mali pesticide use fell from an average of 4.5 to 0.25 kg of active ingredient per ha: Settle and Hama Garba, 2011). In some cases, biological control agents have been introduced where pesticides were not being used at all, or habitat design has led to effective pest and disease management (Royal Society, 2009; Khan *et al.*, 2011). The greater diversity of trees, crops (e.g. beans, fodder shrubs, grasses) and non-cropped habitats has generally helped to reduce run-off and soil erosion, and thus increased groundwater reserves.

Projects across sub-Saharan Africa, where nutrient supply is a key constraint, have used a mix of inorganic fertilisers, organics, composts, legumes, and fertiliser trees and shrubs to improve nutrient availability, in conjunction with conservation tillage to improve soil health. Policy and institutional support has also been important. The Malawi fertiliser subsidy programme is a rare example of a national policy that has led to substantial changes in farm use of fertilisers and the rapid shift of the country from food-deficit to food-exporter (Dorward and Chirwa, 2011). In this case, the importance of both bonding social capital between farmers in groups and linking social capital between national institutions and farmers was critical to rapid adoption.

Some of the best evidence comes from analysis of IPM programmes in Asia and Africa. There are relatively few cross-country evaluations of the effectiveness of IPM (van den Berg, 2004; Tripp *et al.*, 2005; van den Berg and Jiggins, 2007; Pretty, 2008). In a recent paper (Pretty and Bharucha, 2015), we analysed 85 IPM projects from 24 countries of Asia and Africa implemented over a 25-year period (1990–2014) in order to assess outcomes on productivity and reliance on pesticides. Projects had been implemented in Bangladesh, Cambodia, China, India, Indonesia, Japan, Laos, Nepal, Pakistan, Philippines, Sri Lanka, Thailand and Vietnam; and in Burkina Faso, Egypt, Ghana, Kenya, Mali, Malawi, Niger, Senegal, Tanzania, Uganda and Zimbabwe.

We again chose projects as units of analysis as they represented a deliberate attempt to address an existing problem (e.g. a new pest or overuse of a particular compound) or to create a new benefit based on research or knowledge from elsewhere in the field (e.g. new agroecological design). We searched the published literature for evaluations of such projects from Asia and Africa, and identified 85 projects with 115 datasets on combined changes to crop productivity and pesticide use for rice, maize, wheat, sorghum/millet, vegetables, potato/sweet potato, soybean/bean and cotton/tea. The time elapsed from project start to measurement of reported impact varied between one to five years. Across these projects, we estimated that 10–20 million farmers on 10–15 Mha had adopted a variety of effective IPM methods and approaches. The best of these involved substantial redesign of agroecosystems.

In principle, there are four possible trajectories an agroecosystem can take with the implementation of IPM (see Figure 4.2):

i both pesticide use and yields increase (PY);
ii pesticide use increases but yields decline (Py);
iii both pesticide use and yields fall (py);
iv pesticide use declines, but yields increase (pY).

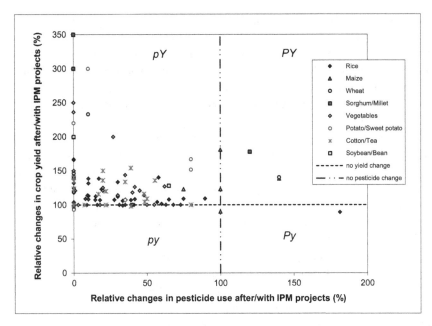

FIGURE 4.2 Impacts of IPM projects and programmes on pesticide use and crop yields (data from 115 crop combinations, 85 projects, 24 countries of Africa and Asia)

The conventional assumption is that pesticide use and yields are positively correlated, suggesting that only trajectories into PY or py are likely. A shift into Py should be against economic rationale, as farmers' returns would be lowered, and thus there should be incentives to change practices. A shift into pY would suggest that current pesticide use is inefficient and could be amended.

Pesticide use data is derived from records of changes to the number of sprays per hectare and/or the amount of active ingredient (a.i.) applied per hectare. Yield data is recorded as kg/ha per crop. Some data is derived from longitudinal studies (where projects were evaluated before and after) or latitudinal studies (where projects are evaluated with and without intervention). The double delta model involves both latitudinal and longitudinal comparisons, and is considered robust (van den Berg and Jiggins, 2007). Evaluations may suffer from placement bias (non-random programme placement through selection of areas or communities easiest to work with or most likely to succeed) and non-random assignment (participation in farmer field schools (FFS) is voluntary and self-selecting). Ideally, there would be exact matching of farmers with IPM/FFS as a treatment and those without, but again it is usually methodologically impossible to control for a range of factors such as farm design, off-farm income, age, access to credit, ownership of mobile phones, access to tarmac roads and irrigation, and education level of family members.

It has not been possible to standardise the time elapsed since introduction of interventions to measured impacts (this varies in these projects from one to five years) or the physical distance between with and without treatments (leaving open the opportunity of farmers learning from one another and reducing the difference between with and without treatment). Again inherent is the potential for selection bias: these are all published studies, which will have involved selection of places, people and projects where there have been outcomes worth reporting (whether positive or negative). With these comments, two key questions are material: is something interesting happening, and are farmers seeing sufficiently different impacts to persist with SI practices?

The mean yield change across projects and crops is an increase of 40.9 per cent, combined with a decline in pesticide use to 30.7 per cent of the original use. A total of 35 of 115 (30 per cent) crop combinations resulted in a transition to zero pesticide use (though none of the projects were formally called organic or involved transitions to organic standards of production); 18 of 115 (16 per cent) crop combinations resulted in no changes to yields but did reduce costs (Figure 4.2).

We found only one Py case (increased pesticide use combined with reduced yield) in the literature, reported by Feder *et al.* (2004). This evaluation considered the impact of FFS training in Indonesian rice cultivation. A number of py (declines in both pesticide use and yield) cases have been reported elsewhere for transitions in industrialised systems typical in Europe following adoption of integrated farming systems. Here, large reductions in pesticide use were accompanied by up to 20 per cent falls in yields (Röling and Wagemakers, 1997). PY contains examples that have involved the adoption of conservation and no-till agriculture, where the reduced tillage improves soil health, reduces erosion and improves surface water quality, but also required increased use of herbicides for weed control. The rapid growth in herbicide use in Argentina and Brazil (see Chapter 2) has followed the widespread adoption of zero-tillage systems, and these show a transition to PY (though this is not inevitable, as there are organic ZT systems in Latin America: Petersen *et al.*, 2000). The majority of cases reported here show transitions to the pY sector (pesticide use falls, yields increase).

While pesticide reductions with IPM may be expected, yield increases induced by IPM are more complex. IPM may, for example, reduce the incidence of severe-loss years, though not increase yields in a normal year, thus increasing mean production across years. Many IPM projects involve interventions focused on more than just pest management (e.g. push–pull systems where legumes fix soil nitrogen as well as help manage parasitoids and depress weeds), or if they involve a significant component of farmer training (e.g. through FFS), then farmers' capabilities at innovating in a number of areas of their agroecosystems will have increased, such as in soil and water management (Settle *et al.*, 2014).

None of this is bad. Yield improvements may also be accompanied by new income streams that open up with increased flows of ecosystem goods and services. Push–pull IPM systems, for example, may provide improved livestock feed, enabling smallholders to diversify into dairying and poultry. This in turn provides increased manure for farms (helping soil health and yields), and also translates into increased income and better nutrition. Farmers may also be able to invest cash savings made from reducing pesticide use into better seeds and fertilisers, both of which would increase yields. These arrays of improvements are good examples of some of the synergistic cycles which can result from sustainable intensification.

Carbon benefits of sustainable intensification

As we noted earlier, it is clear that both emission reductions and sink growth will be necessary for mitigation of current climate change trends. A source is any process or activity that releases a greenhouse gas, or aerosol or a precursor of a greenhouse gas into the atmosphere, whereas a sink is any mechanism that removes these from the atmosphere. Carbon sequestration is defined as the capture and secure storage of carbon that would otherwise be emitted to or remain in the atmosphere. Soils already contain more carbon than the atmosphere and terrestrial vegetation combined: 1500 Pg in the top metre, compared with 760 Pg in the atmosphere and 560 Pg in all biota (Lal, 2004).

Agricultural systems emit carbon through the direct use of fossil fuels in food production, the indirect use of embodied energy in inputs that are energy intensive to manufacture, and the cultivation of soils and/or soil erosion resulting in the loss of soil organic matter. Agriculture also contributes to climate change through the emissions of methane from irrigated rice systems and ruminant livestock. The direct effects of land use and land-use change (including forest loss) have led to a net emission of large quantities of carbon. On the other hand, agriculture is also an accumulator of carbon when organic matter is accumulated in the soil, and when above-ground biomass acts either as a permanent sink or is used as an energy source that substitutes for fossil fuels and thus avoids carbon emissions. There are three main mechanisms and 19 practices (Table 4.3) by which positive actions can be taken by farmers:

i   increasing carbon sinks in soil organic matter and above-ground biomass;
ii avoiding carbon dioxide or other greenhouse gas emissions from farms by reducing direct and indirect energy use;
iii increasing renewable energy production from biomass that either substitutes for consumption of fossil fuels or replacing inefficient burning of fuelwood or crop residues, and so avoids carbon emissions.

Conservation agriculture involving low or zero tillage has emerged as a SI system with considerable potential for carbon sequestration. Adoption of CA is important for restoring soil quality by reducing the loss of soil organic carbon (SOC) stock and improving soil microbial biomass carbon. Continuous use of CA in China has increased soil organic matter in the surface layer at the rate of 0.01 per cent per year (Li *et al.*, 2007; He *et al.*, 2011). Such an increase is often observed after long-term (>10 years) application of CA, with the magnitude of increase as much as >2.0 g kg^{-1} in the 0–10 cm layer (Li *et al.*, 2007; He *et al.*, 2009, 2011). Increase in SOC and biotic activity reduces soil bulk density, improves pore size distribution and increases soil fertility. Improvement in soil structure under CA mitigates soil degradation and reduces the loss of cropland, decreases emissions of methane and creates a positive soil C budget particularly in highly erodible soils (Pan *et al.*, 2009).

In another meta-analysis (Li *et al.*, 2016), we were able to compare paired data from China on yields and soil organic carbon concentration (0–20cm soil depth) (Figure 4.3). In 34 datasets, 26 (77 per cent) showed that CA increased both yields and SOC (CY quadrant) as compared with traditional inversion tillage, and there were no cases of both reductions in yield and SOC. In all the datasets, CA showed increased SOC, and particularly in 0–10cm soil depth, the mean increase could be >3g kg^{-1}.

TABLE 4.3 Mechanisms for increasing carbon sinks and reducing CO_2 and other greenhouse gas emissions in agricultural systems

---

Mechanism A. Increase carbon sinks in soil organic matter and above-ground biomass
- Replace inversion ploughing with conservation- and zero-tillage systems
- Adopt mixed rotations with cover crops and green manures to increase biomass additions to soil
- Adopt agroforestry in cropping systems to increase above-ground standing biomass
- Minimise summer fallows and periods with no ground cover to maintain soil organic matter stocks
- Use soil conservation measures to avoid soil erosion and loss of soil organic matter
- Apply composts and manures to increase soil organic matter stocks
- Improve pasture/rangelands through grazing, vegetation and fire management both to reduce degradation and increase soil organic matter
- Cultivate perennial grasses (60–80 per cent of biomass below ground) rather than annuals (20 per cent below ground)
- Restore and protect agricultural wetlands, convert marginal agricultural land to woodlands to increase standing biomass of carbon.

Mechanism B. Reduce direct and indirect energy use to avoid greenhouse gas emissions (CO_2, CH_4 and N_2O)
- Conserve fuel and reduce machinery use to avoid fossil fuel consumption
- Use conservation or zero tillage to reduce CO_2 emissions from soils
- Adopt grass-based grazing systems to reduce methane emissions from ruminant livestock
- Use composting to reduce manure methane emissions, substitute biofuel for fossil fuel consumption
- Reduce the use of inorganic N fertilisers (as manufacturing is highly energy intensive), and adopt targeted- and slow release fertilisers
- Use IPM to reduce pesticide use (avoiding indirect energy consumption).

Mechanism C. Increase biomass-based renewable energy production to avoid carbon emissions
- Cultivate annual crops for biofuel production such as ethanol from maize and sugar cane
- Cultivate annual and perennial crops, such as grasses and coppiced trees, for combustion and electricity generation, with crops replanted each cycle for continued energy production
- Use biogas digesters to produce methane, so substituting for fossil fuel sources
- Use improved cook-stoves to increase efficiency of biomass fuels.

The SOC sink capacity of China's farmland could be increased by 70–250 kg C ha^{-1} yr^{-1} through conversion to CA systems (Table 4.4). Combined with reductions in fossil fuel consumption associated with reduced ploughing, increases of SOC in CA are equivalent to 89–269 kg C ha^{-1}yr^{-1}, amounting to an annual increase of 7.2–21.8 Mt C (0.02–0.07 Gt CO_2e) from 60 per cent of China's farmland. Adoption of CA would thus produce a contribution to climate change mitigation, though not as much as had been proposed in the UNEP Emissions Gap Report (UNEP, 2014). Our estimate here of the CO_2 emission reduction through adoption of CA (0.04–0.13 Gt CO_2e yr^{-1}) would amount to 1–6 per cent of total global emissions from adoption of CA. This would contribute to the COP-21

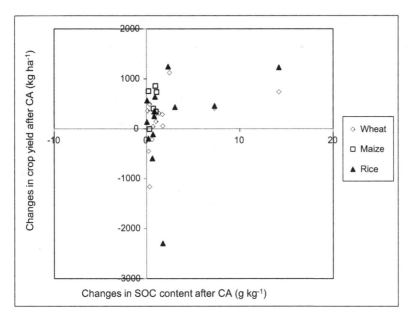

FIGURE 4.3 Relationship between changes in yield and soil organic carbon under conservation agriculture, China

Notes

CY – top right: both organic carbon and yields increase;

Cy – bottom right: organic carbon increases but yields decline;

cy – bottom left: both organic carbon and yields fall;

cY – top left: organic carbon declines, but yields increase.

Soil organic carbon (0–20cm soil depth) and crops yields were measured at the end of each experiment

TABLE 4.4 The effect of conservation agriculture on soil carbon improvement in China relative to conventional practices

Factor	Impact on carbon balance (conservative annual values)
Fossil fuel use reduction (20kgha⁻¹)	~18.8 kg C–e/ha
Soil C sequestration	70–250 kg C/ha
Total	88.8–268.8 kg C/ha

Sources: Dong *et al.* (2009); DAMM (2012); Liu *et al.* (2014).

Note

Assuming 60 per cent (based on CA's adoption area in the main countries, such as Australia, Canada, Brazil, which have developed CA system) Chinese farmland (total 135 Mha in 2013) is converted to CA, estimate of the total potential of C sequestration in China cropland can be increased by 7.2 to 21.8 Mt C (0.02–0.07 Gt CO_2–e)/year.

recommendation regarding societal value of carbon and the need for targets for carbon sequestration in soils (Lal *et al.*, 2014).

TABLE 4.5 Numbers of farmers by country (selected countries, over 1M for developing countries)

Asia		Africa		Latin America		Industrialised countries	
Country	Number of farmers (million)	Country	Number of farmers (million)	Country	Number of farmers (million)	Country	Number of farmers (million)
Afghanistan	3.0	Angola	1.0	Brazil	5.0	Australia	0.14
Bangladesh	15.0	Côte d'Ivoire	1.1	Colombia	2.0	Canada	0.24
China	200.0	DR Congo	4.5	Mexico	5.5	France	0.67
India	138.0	Egypt	4.5	Peru	2.3	Italy	2.6
Indonesia	25.0	Ethiopia	10.7			Japan	3.1
Myanmar	5.4	Ghana	1.8			Poland	2.9
Nepal	3.4	Kenya	2.7			South Korea	3.2
Pakistan	6.6	Madagascar	2.4			UK	0.22
Philippines	4.8	Malawi	2.7			USA	2.2
Sri Lanka	3.3	Morocco	1.5				
Thailand	5.8	Mozambique	3.0				
Vietnam	10.7	Rwanda	1.7				
		South Africa	1.1				
Azerbaijan	1.3	Uganda	3.8				
Iran	4.0	Tanzania	4.9				
Kyrgyzstan	1.1	Zambia	1.3				
Turkey	3.0						
Yemen	1.5						

Source: Lowder *et al.* (2016).

The primary focus: small farms

There are some 570 million farms worldwide, 75 per cent of which are small farms of less than 2 ha (Lowder *et al.*, 2016). They make up 12 per cent of agricultural area, and produce 70 per cent of world food. It is clear where sustainable intensification will have to be effective. Some 74 per cent of all farms are in Asia (of which 35 per cent are in China, and 24 per cent in India), 9 per cent in sub-Saharan Africa, 7 per cent in Central Europe and Central Asia, 3 per cent in Latin America and the Caribbean, 3 per cent in the Middle East and North Africa, and only 4 per cent in the industrialised and affluent countries (Table 4.5).

We now turn to the SI innovations and redesigns that have been developed and promoted in developing countries. The evidence is quite clear, and may appear to be counter-intuitive. Developing countries and their farmers are leading industrialised farmers in the implementation of sustainable intensification fitted to the needs and opportunities of their social and ecological systems.

5

SUSTAINABLE INTENSIFICATION ON SMALLER FARMS IN DEVELOPING COUNTRIES

Across the world, progress towards sustainable intensification – predominantly on small farms – is taking place in the context of two priorities. The first is the pressing need to implement sustainable development goals for poverty reduction, improved livelihoods and better nutrition by building more productive and sustainable systems of smallholder agriculture. The second key consideration is a deepening pull from global markets, which is intensifying transitions to larger enterprises catering to global commodity supply chains and high-value crops. We do not address the implications of the latter set of developments in this book, focusing instead on the remarkable progress being made across small farms growing both food and cash crops.

There is not the space in any one book to detail all the changes towards sustainable intensification that have been made on small farms in developing countries. This is a sign of success. We thus have chosen examples especially to illustrate key principles of redesign and components of substitution, focusing on sustainable intensification practices that have expanded to scale, usually resulting in improvements to the livelihoods of many farmers and the environments of many farms. These illustrate how new thinking and practices are already producing significant impacts, many now occurring at scale. We will also show further examples in Chapter 7 where social capital has been critical for redesign.

Crop variety improvements

Improved crop varieties, particularly those which yield better and/or express resistance to certain pests and diseases, are at the forefront of agricultural intensification across developing countries. Varietal improvements during the green revolutions resulted in remarkable improvements for key staples: wheat (208 per cent), paddy rice (109 per cent), maize (157 per cent), potato (78 per cent) and cassava (36 per cent) between 1960 and 2000 (Pingali and Raney, 2005). Conventional plant breeding was at the heart of these improvements – with plants from different genetic backgrounds crossed, and then selected when they expressed desirable traits. As we will outline later, the public sector was an important driver of these improvements. Indeed, they could not have occurred without policy support at all scales, from local extension agents to international research consortia, all supported by public funds.

From the 1990s onwards, varietal improvement has also focused on the methods of biotechnology and genetic modification (GM). GM crops have seen a remarkable spread. First commercialised in 1996, there has been a 100-fold increase in global hectarage. Worldwide, some 18 million farms in 26 countries cultivate them (ISAAA, 2016). They are as important on small farms in developing countries as they are in industrialised countries. Just over half of the global hectarage under GM crops was in Latin America, Asia and Africa, the primary crops being soybean, cotton, maize and canola (oil seed rape). In a 2006 review, Raney concluded that "the economic evidence available to date does not support the widely held perception that transgenic crops benefit only large farms; on the contrary the technology may be pro-poor. Nor does the available evidence support the fear that multinational biotechnology firms are capturing all the economic value created by transgenic crops".

This gene revolution has proceeded along a markedly different path than previous waves of varietal improvement. Private investment has been predominant (Gray and Dayananda, 2014), with a relatively small number of multinational bioscience corporations spending billions on research and development. Yet, state policy remains important, and not just for regulation of seeds and markets. China, Brazil and India lead the world in public investment in agricultural biotechnology, and many other developing countries have increasing capacity to adopt and adapt innovations developed elsewhere.

Varietal improvement is one of the most important ways to improve productivity, particularly of staples like maize, rice and wheat (FAO, 2016c). It can also improve the resilience of the whole farming system. Adopting high-yielding varieties can also open up opportunities for adding new system elements. On the coastal polders of Bangladesh, farmers who have adopted high-yielding rice have been able to add a second, dry season crop. After harvesting rice early, farmers have planted okra, sesame, mungbean and sunflower during the dry season.

Yet, sustainable intensification means new priorities for plant breeding. First, new varieties need to be better adapted to ecosystem-based agriculture, and focus on improved nutritional content, better resource-use efficiency and the reduction of GHG emissions (Bouis, 1996; Tilman et al., 2002). Work on this is in progress, with proposals for a breeding programme for wheat cultivars suited to zero tillage. Among traits that have been identified are more rapid establishment and faster root growth (Kohli and Fraschina, 2009). Other research institutions are responding to increasing water scarcities by identifying wheat genotypes with higher water-use efficiency (Solh and van Ginkel, 2014).

The Bangladesh Agricultural Research Institute and CIMMYT (International Maize and Wheat Improvement Center) have adapted and promoted drill-seeders originally developed for wheat, so that they can be used to sow maize and rice without tillage. In north-west Bangladesh, farmers using these seeders obtained rice yields similar to those of transplanted rice, but using less water and labour, and were able to harvest the crop two weeks earlier. A study in Bangladesh compared yields and profitability under ploughing and zero tillage. With permanent bed planting of maize, the combined productivity of rice and maize was 13.8 tonnes per ha, compared with 12.5 tonnes on tilled land. The annual cost of rice–maize production on permanent beds was US$1,530 per ha, compared with US$1,680 under conventional tillage (Gathala *et al.*, 2014).

Hybrid maize requires large amounts of nitrogen to produce high yields. But Bangladesh's reserves of natural gas, which is used to produce urea fertiliser, are finite and non-renewable. One promising solution to soil nutrient depletion is the application of poultry manure, which is becoming abundant – Bangladesh's poultry sector now produces about 1.6 Mt of manure every year (Ali *et al.*, 2009). Good maize yields have been obtained by replacing with poultry manure 25 per cent of the mineral fertiliser normally applied. Soil nitrogen can also be partially replenished by growing legumes, such as mungbeans, after the maize harvest. In tropical monsoon climates, a summer mungbean crop also mops up residual nitrogen and prevents nitrate pollution of aquifers.

Planting short duration rice varieties allows farmers to plant maize earlier. However, those rice varieties produce lower yields, and farmers are generally unwilling to sacrifice the production of their main food crop. The Bangladesh Rice Research Institute is thus developing higher-yielding, shorter duration Aman rice varieties. The future of sustainable rice–maize farming in South Asia also hinges on the development of high-yielding maize hybrids that mature quickly and tolerate both waterlogging and drought.

The development of these varieties requires plant breeders to have continual access to the widest possible sources of desirable traits, which are found in cereal accessions held in germplasm collections, in wild relatives and in landraces in farmers' fields. Second, varietal development needs to be matched by seed systems that ensure the rapid multiplication of the seed and its supply to smallholder farmers. Seed production is especially critical for hybrids of cross-pollinated crops such as maize. In many countries, the lack of such systems has prevented farmers from adopting new varieties and improved farming practices. Many farmers continue to use seed from harvests in their fields. This 'saved seed' is often of poor quality, and its repeated use can result in low yields.

Third, there is scope for increasing the share of perennials in the global crop mix. Annuals currently cover around 70 per cent of global cropped area, but perennials offer several advantages for ecosystem services and cost-savings to farmers (Dewar, 2007; Glover and Reganold, 2010; Kell, 2011; Crews and Brookes, 2014). Beneficial traits may include the ability to be grown on resource-poor and 'marginal' lands and the ability to sustain more production per unit of land than most annual crops grown on fertile lands. Varietal development is also needed in order to produce perennial alternatives that express desired traits – larger seed size, stronger stems, improved palatability and higher seed yield (Glover and Reganold, 2010). Breeding plants with deeper and bushy root systems may also offer improved soil structure, water and carbon sequestration, nutrient retention and higher yields (Kell, 2011). Estimates suggest that $5–10$ kg/m^2 ($50–100$ t/ha) could be sequestered through increased root mass, resulting in globally significant sequestration of atmospheric carbon.

Finally, there have also been calls for increased attention to orphan crops in developing countries (Bioversity International, 2016). These are varieties which are "valued culturally, often adapted to harsh environments, nutritious, and diverse in terms of their genetic, agroclimatic, and economic niches" (Naylor *et al.*, 2004). They are important for nutritional security in the world's most disadvantaged regions. Many orphan varieties and legumes are also coming to attention as a result of international research collaborations and advances in sequencing and genotyping technologies (Bohra *et al.*, 2014).

Sweet potato is a particularly important orphan crop in sub-Saharan Africa, where it underpins farm-level food security as an important source of starch in addition to calcium and riboflavin. In Uganda, yields for sweet potato are around 4 t/ha (compared with an average global yield of 14 t/ha) (Naylor *et al.*, 2004). Conventional breeding for sweet potato is relatively slow because it is a vegetatively propagated perennial; biotechnology-based approaches on the other hand could confer pest resistance to weevils and viruses and improved starch/dry matter ratios, and it has been estimated that effective resistance could raise yields to 7 t/ha. Transfer of the technology across half the sweet potato area in sub-Saharan Africa would result in a gross annual benefit of US$121 million (Naylor *et al.*, 2004). In Kenya, it has been estimated that effective resistance to viruses and weevils could result in income increases of 28–39 per cent (Qaim, 1999). Other approaches to sweet potato improvements have centred on conventional breeding to improve vitamin A content and to shorten time to harvest (Mwanga and Ssemakula, 2011). Emerging evidence demonstrates the value of participatory varietal development. Other important orphan crops include tef and white pea bean in Ethiopia and cassava in Ghana. All have shown the remarkable potential of participatory breeding (Assefa *et al.*, 2011, 2014; Manu-Aduening *et al.*, 2014).

Integrated pest management

Integrated Pest Management (IPM) consists of a toolbox of management decisions and interventions designed to combine the use of targeted compounds, agronomic and biological techniques to control crop pests. Contemporary IPM systems have had several traditional and modern precursors, as farmers and agricultural scientists have long struggled to contain pest populations where inorganic compounds were unavailable (Birch *et al.*, 2011). Contemporary IPM emerged after World War II following the recognition that indiscriminate use of insecticide would be ecologically problematic. Early supervised and integrated control projects were first developed for alfalfa caterpillar and spotted alfalfa aphid (Ehler, 2006). Since then, it has been stated that IPM has become the dominant crop protection paradigm, yet its adoption remains low (Parsa *et al.*, 2014). As indicated earlier, it has not yet led to a reduction in total pesticide use, nor has it yet eliminated negative externalities. There are also an increasing number of new invasive pests and diseases being discovered, as transfer of species in a globalised world has become easier, and changes in climate and weather patterns have driven shifts in pest and pathogen ranges.

Preventing crop losses to pests is essential for meeting sustainable development goals around hunger and nutritional security. Worldwide, crops lost to pests represent the equivalent of food required to feed one billion people (Birch *et al.*, 2011). Over coming decades, pest damage may worsen due to the response of pest species to global climate change and the distribution of pests and pathogens will change (Bebber *et al.*, 2013). Resistance poses a continuing challenge. Since the discovery of triazine resistance in common groundsel in the late 1960s, herbicide resistance in weeds has grown rapidly. Globally, some 220 weed species have evolved herbicide resistance, posing a particular challenge to cereals and rice (Heap, 2014). Five species (*Avena* spp., *Lolium* spp., *Phalaris* spp., *Setaria* spp. and *Alopecurus myosuroides*) infest over 25 Mha of cereal crops globally (Heap, 2014). For insect pests, over 600 cases of insecticide resistance have become apparent since the 1950s (Pretty, 2005; Head and Savinelli, 2008).

Synthetic pesticides have had mixed impacts. In the short-term, they prevent significant crop losses. However, they threaten human and ecosystem health and effectiveness declines as resistance develops (Pretty, 2005). Thus, over-reliance on agrichemicals as a sole form of plant protection is not sustainable (Shaner, 2014). Moreover, with the development of poorly regulated generics, especially in developing countries, cheap alternatives which do not meet international quality standards 'lock-in' the use of obsolete pesticides (Popp *et al.*, 2013) and result in associated risks for ecosystems and human health (de Bon *et al.*, 2014).

Thus, complementary and alternative modes of pest control, relying on pest ecologies, have been gaining increasing attention as part of a suite of diversified strategies to control pest populations and increase crop resilience to infestation. The use of on- and off-farm biodiversity is an overarching principle in new forms of pest management. A key principle is that biodiverse agroecosystems demonstrate less pest damage and more natural pest enemies than non-biodiverse ones (Letourneau *et al.*, 2011).

Effective IPM centres on the principle of deploying multiple complementary methods for pest, weed and disease control. Both social and human capital are important. IPM is knowledge-intensive, involving a set of decisions for coordinating multiple tactics of different classes of pests (Ehler, 2006). Farmers need to monitor pests and natural enemies, understand thresholds for decisions and be competent in a range of different methods from pesticide product management/substitution to whole agroecosystem redesign. This broad range of options also allows for many interpretations of IPM (Parsa *et al.*, 2014; Gadanakis *et al.*, 2015).

TABLE 5.1 ESR options for integrated pest management

Type	Examples of application
Efficiency	
1 Management of application of pesticides	• Targeted spraying • Threshold spraying prompted by decision-making derived from observation/data on pest, disease or weed incidence
Substitution	
2a Substitution of pesticidal products with other compounds	• Synthetic pesticide with high toxicity substituted by another product with low toxicity • Use of agrobiologicals or biopesticides (e.g. derived from neem)
2b Releases of antagonists, predators or parasites to disrupt or reduce pest populations	• Sterile breeding of male pest insects to disrupt mating success at population level • Identification and deliberate release of parasitoids or predators to control pest populations
2c Deployment of pheromone compounds to move or trap pests	• Sticky and pheromone traps for pest capture
2d Crop and livestock breeding	• Deliberate introduction of resistance or other traits into new varieties or breeds (e.g. recent use of genetic modification for insect resistance and/or herbicide tolerance)
Redesign	
3 Agroecological system and habitat redesign	• Seed and seedbed preparation • Deliberate use of domesticated or wild crops/plants to push–pull pests, predators and parasites • Use of crop rotations and multiple cropping to limit pest, disease and weed carry-over across seasons or viability within seasons • Adding host-free periods into rotations • Adding stakes to fields for bird perches

Source: From Pretty and Bharucha (2014).

Overall, IPM approaches can be classified into six main types (Table 5.1). These vary along a spectrum from targeted or changed use of pesticide compounds to habitat and agroecological design. In only very rare cases, such as the aerial release of the parasitic wasp, *Epidinocarsis lopezi*, to control cassava mealybug in West and Central Africa (Herren *et al.*, 1987; Neuenschwander, 2001), can IPM be implemented without farmer engagement in locally relevant and effective methods for IPM. In the past 25 years, there has been a substantial increase in understanding on how to increase farmers' knowledge so that they are able to cultivate and raise crops and livestock whilst reducing or eliminating pesticides. The principles are to find appropriate ways of building the natural, social and human capital in ecosystems such that these provide sufficient services to maintain or increase agricultural productivity whilst reducing or eliminating environmental harm.

Social capital matters more and more because, over time, IPM strategies have transitioned from individual field-based practice to coordinated, community-scale decision-making covering wider landscapes (Brewer and Goodell, 2012). While this improves the effectiveness of pest control, it presents a significant obstacle to wider adoption (notably in developing country contexts) by presenting a collective-action dilemma (Pretty, 2003). Farmer field schools (FFS) play a central role in training farmers in IPM, and have been shown to result in improved outcomes (van den Berg and Jiggins, 2007; Pretty *et al.*, 2011a; Settle *et al.*, 2014).

Farmer field schools, beginning in the 1980s (Kenmore *et al.*, 1984), are amongst the most significant innovations for the development and spread of IPM. These build both social and human capital, bringing education, co-learning and experiential learning so that farmers' expertise is improved to provide resilience to current and future challenges in agriculture. FFS are not just an extension method: they increase knowledge of agroecology, problem-solving skills, group building and political strength. FFS have also been complemented by modern methods of extension involving video, radio, market stalls, pop-ups and songs (Bentley, 2009). These can be particularly effective where there are simple messages or heuristics that research has shown will be effective if adopted. FFS have now been used in 90 countries (Braun and Duveskog, 2009; FAO, 2016d; Ketelaar *et al.*, 2018), and some 12 million farmers have graduated from FFS, including 650,000 in Bangladesh, 250,000 in India, 930,000 in Vietnam, 1.1 million in Indonesia, 500,000 in the Philippines and 90,000 in Cambodia. Now some 20,000 FFS graduates are running FFS for other farmers, having graduated from farmer to expert trainer. We will discuss the remarkable social innovation of farmer field schools more in Chapter 7.

One of the most effective and rapidly adopted IPM systems is the *push–pull* system, which is yielding notable successes in monocropped cereal systems (Cook *et al.*, 2007; Royal Society, 2009; ICIPE, 2013). Khan *et al.* (2017) estimate that across Kenya, Uganda, Tanzania and Ethiopia, climate-smart push–pull systems are used across 130,000 small farms. These have been deployed with great effect against *Striga* weed and stem-borer infestations in maize, millet and sorghum (Khan *et al.*, 2014). Interplanting of the leguminous forage crop *Desmodium* suppresses *Striga* and repels stem-borer moths while attracting their natural enemies; planting *Napier* grass as a border crop attracts stem-borer moths. Positive externalities include nitrogen fixation by *Desmodium* and the provision by new plants of high quality animal fodder. This has enabled farmers to diversify into dairy and poultry production, which in turn has increased the availability of animal manure for application on fields. Across

East Africa, push–pull systems have resulted in yield increases of 23 per cent, and 75 per cent decreases in pesticide use.

The push–pull technology provides both economic and environmental benefits, ranging from guaranteed increased yields by at least 3 t/ha, and food sufficiency achieved through the control of the major abiotic and biotic constraints through soil fertility improvements and enhancement of agroecosystem integrity and resilience. Further development and adaptation of the technology to fit a wider range of farm typologies and farmer practices are extending its benefits to more farmers. The new drought-tolerant trap plants and intercrops effectively control stem-borers and *Striga* weeds. As a result, yields of maize and sorghum have increased significantly, with up to threefold increase over control plots. The new companion plants have also provided ample, good quality livestock fodder, producing enough to allow farmers to make hay for the dry season. Better quality of fodder for dairy animals has increased milk yields by at least 2 litres per day. In all, the system gives significantly higher economic returns to the farmer than conventional farmer practices

The on-farm uptake of conventional and climate-adapted push–pull by large numbers of farmers in eastern Africa has confirmed that push–pull has been widely accepted, and has significant impacts on food security, human and animal health, soil fertility, income generation, empowerment of women, conservation of agro-biodiversity and provision of agroecosystem services. Climate-smart push–pull is spreading these benefits in a wider range of farming systems, conferring the benefits to additional crops and agroecosystems.

Conservation agriculture

Conservation Agriculture (CA) consists of a variety of measures to minimise soil disturbance, maintain soil cover and rotate crops. The aim is to mitigate soil erosion, improve water-holding capacity and increase soil organic matter to improve soil health and boost crop yields. A key feature is the revision or reduction of soil disturbance through tilling. Zero tillage involves no ploughing prior to sowing. This seeks to maintain an optimum environment in the root zone in terms of water availability, soil structure and biotic activity (Kassam *et al.*, 2009).

CA evolved in part as a response to the severe soil erosion that devastated the US Midwest in the 1930s. Currently, CA systems are practised successfully across a range of agroecological conditions, soil types and farm sizes (Derpsch *et al.*, 2010). At present, CA practices cover just over 8 per cent of global arable cropland, but are estimated to be spreading globally by some 6 million hectares per year to a total of 180 Mha in 2017. Adoption varies greatly by region. CA covers some 69 per cent of arable cropland in Australia and New Zealand, 57 per cent of arable cropland across South America and 15 per cent in North America. By contrast, adoption across Europe and Africa is low (covering only 1 per cent of arable cropland in each of the two continents) (Kassam *et al.*, 2014; Jat *et al.*, 2014).

CA systems result in yield improvements of 20–120 per cent compared with conventional tillage systems (Kassam *et al.*, 2009). Resource efficiency is a key feature.

CA systems often reduce the need for fertiliser application over time, produce lower run-off and build resilience to pest and disease. All these result in significant savings, which, combined with yield increases, may translate to significant financial benefits for farmers relative to conventional ploughing practice (Sorrenson, 1997). A recent European Conference on Green Carbon highlighted the importance of soil carbon content as a marker and enabler of sustainability in agroecosystems, and concluded that CA presents a good strategy to maintain and improve soil carbon levels. Comparisons between conventional tillage and reduced or no-till systems have found higher soil organic carbon, lower emissions and improved soil quality under the latter (Brenna et al., 2013 in Italy; Franzluebbers, 2010 for the south-eastern USA; Spargo et al., 2008 in Virginia), especially on certain types of soils and with the addition of cover cropping.

While CA offers much potential for sustainable intensification, scientific debate highlights a number of difficulties and contentions. Each component in the CA package requires interpretation and the applicability and scalability of CA to smallholder systems has been questioned, especially in developing countries (Giller et al., 2009, 2011; Stevenson et al., 2014). Nevertheless, some case studies show remarkable social–ecological outcomes, including higher yields, as well as reduced soil loss, increased soil carbon content, improved soil structure and water productivity compared with conventional methods (Marongwe et al., 2011). Collectively, recent evidence from the African Union (Pretty et al., 2011a, 2011b) shows that the adoption of CA principles had led to improvements on 26,000 ha, with a mean yield increase ratio of 2.20 and annual net multiplicative yield increases in food production of some 11,000 tonnes/year. In addition, a number of positive externalities, with cost-saving or income-boosting effects, were also reported, including reduced soil erosion, increased resilience to climate-related shocks, increased soil carbon, improved water productivity, reduced debt, livelihood diversification and improved household-level food security (Silici et al., 2011; Marongwe et al., 2011; Owenya et al., 2011). In Vietnam, no-till strategies allowed households headed by women and the elderly to participate in growing potatoes, previously abandoned in favour of less labour-intensive crops. The need for weeding is reduced by adequate mulching, and tubers are harvested simply by moving aside the rice straw that covers them over.

Going forward, the evidence base will need to be strengthened (Brouder and Gomez-MacPherson, 2014). Meta-analyses and reviews across cases show that the evidence on yield impacts and carbon sequestration potential is mixed (Stevenson et al., 2014). This may, in part, reflect the context-sensitivity of CA, where outcomes depend on the precise combination of practices used, and differ by crop type. There is some evidence that zero tillage may result in yield penalties in the short term (Brouder and Gomez-MacPherson, 2014). This may hinder adoption amongst smallholders, particularly those navigating pressing concerns around food and livelihood security (Giller et al., 2011). Other studies find yield increases to be stable over time. For example, Derpsch (2008) compared conventional and CA systems over a decadal scale, and found yield decreases over the period in the former, and yield increases over the period under CA, in addition to lower use of inputs. There is also a need for more evidence on the implications of improved land management across agricultural sectors and agroecosystems (e.g. Morgan *et al.*, 2010), using appropriate soil sampling strategies (Baker *et al.*, 2007) and standardised methodologies (Derpsch, 2014).

A recent analysis of CA uptake in Paraguay, Argentina, Brazil and Uruguay has also shown that small farmers have not adopted CA in large numbers (Derpsch *et al.*, 2016). Large- and medium-scale farmers with tractors have made the transition, and few will return to conventional tillage. But small farmers relying on animal traction or manual farming have not stayed with CA: many have disadopted. Although there are clear technological reasons for these differences, it would appear that the failings lie in the transfer-of-technology approach. CA should be knowledge-based and learning intensive, with farmers capable of adapting and changing. These skills and capacities had not yet been built up amongst smaller farmers.

Conservation agriculture in Kazakhstan

In the spring of 2012, as farmers across the semi-arid steppes of northern Kazakhstan were sowing their annual wheat crop, the region entered one of its worst droughts on record. In many areas, no rain fell between April and September and daily summer temperatures rose several degrees above normal (CIMMYT, 2013). That year, many farmers lost their entire crop and Kazakhstan's wheat harvest, which had reached 23 Mt in 2011, fell to less than 10 Mt. Some farmers, however, did not lose their crops. They were among the growing number of Kazakhstani wheat growers who had adopted conservation agriculture, comprising zero tillage, retention of crop residues on the soil surface and crop rotation. Those practices have increased levels of soil organic carbon and improved soil structure in their fields, allowing better infiltration and conservation of moisture captured from melting winter snow (Nurbekov *et al.*, 2014). As a result, some farmers in Kostanay province have achieved yields of 2 tonnes per hectare, almost double the national average of recent years (CIMMYT, 2013).

Some 2 million of Kazakhstan's 19 Mha of cropland are under full conservation agriculture. On 9.3 Mha, farmers have adopted minimal tillage, which uses narrow chisel ploughs at shallow depths (Karabayev *et al.*, 2014). State policy actively promotes conservation agriculture, and the top priority in agricultural research is the development and dissemination of water-saving technologies. In 2011, Kazakhstan introduced subsidies on ca equipment that are three to four times higher than those on conventional technologies (Nurbekov *et al.*, 2014). Government support has encouraged farmers in northern Kazakhstan to invest an estimated US$200 million to equip their farms with zero-tillage machinery. CIMMYT and FAO, together with Kazakh scientists and farmers, have launched a programme to introduce conservation agriculture in rainfed areas, and raised-bed planting of wheat under irrigation in the south of the country (Karabayev and Suleimenov, 2010). Trials in the north showed zero-tilled land produced wheat yields 25 per cent higher than ploughed land, while labour costs were reduced by 40 per cent and fuel costs by 70 per cent and demonstrated the advantages of growing oats in summer instead of leaving land fallow. With an oat crop, the total grain output from the same area of land increased by 37 per cent, while soil erosion was much reduced.

Kazakhstan now ranks among the world's leading adopters of zero tillage, mainly driven by high adoption rates on large farms and those with rich black soils, where high returns provide the capital needed for investment. Many farmers have found that combining zero tillage with permanent soil cover also helps to suppress weeds. Retaining crop residues increases the availability of water to the wheat crop and reducing or even eliminating erosion, thus addressing desertification and land degradation that costs Central Asian countries an estimated $2.5 billion annually. On-farm research has found that the use of residues to capture snow, along with zero tillage, can increase yields by 58 per cent. Farmers are taking advantage of available rainfall to grow oats, sunflower and canola, and further studies have shown the potential for other rotational crops, including field peas, lentils, buckwheat and flax (Karabayev, 2012). Further increases will be possible with the development of high-yielding wheat varieties better suited to zero tillage and the north's harsh winters and increasingly hot summers.

Fertiliser trees for agroforestry

Agroforestry is a form of intercropping combining perennial trees or shrubs with annual herbaceous crops. Globally, just under half of all agricultural land now has over 10 per cent tree cover, 27 per cent of agricultural land has over 20 per cent tree cover and just over 7 per cent of global agricultural land has over 50 per cent tree cover (Zomer *et al.*, 2009). Not all tree cover necessarily represents agroforestry, but these figures mean that, contrary to early assumptions about trade-offs between cultivated land and tree cover, trees are already integral to most agricultural landscapes. The exceptions are North Africa and West Asia.

A wide variety of agroforestry systems are practised worldwide. Different forms result in varying tree densities, and different mixes between trees, annuals and (sometimes) livestock. Lorenz and Lal (2014) estimate that tree intercropping worldwide covers some 700 Mha, multi-strata systems 100 Mha, protective systems 300 Mha, silvopasture 450 Mha and tree woodlots 50 Mha. Over 100 distinct agroforestry systems have been recorded (Atangana *et al.*, 2014) and modern agroforestry systems have a number of precursors, such as traditional systems across India (Murthy *et al.*, 2013), the Sahel (Garrity *et al.*, 2010) or the Javanese systems of *pekarangan* (Christanty *et al.*, 1986).

Trees bring a variety of social–ecological benefits to farms and surrounding landscapes. It has been estimated that each additional tree planted in an agroforestry system results in an average value of $1.40/year through improved soil fertility, fodder, fruit, firewood and other produce. This would result in additional production value of at least $56 ha/year, and a total production value of $280 million (Larwanou and Adam, 2008). In India, Murthy *et al.* (2013) review carbon sequestration within agrisilvicultural and silvipastoral systems and estimate that these could hold a soil carbon stock of 390 Mt C (excluding soil organic carbon stocks). Ickowitz *et al.* (2013) found that in 21 African countries, dietary diversity amongst children increases with tree cover after controlling for relevant confounding variables. Under Evergreen Agriculture systems, trees are planted within fields of annual crops, and replenish soil fertility, provide food, fodder, timber and fuelwood. Overall, this portfolio of benefits provides "an overall value greater than that of the annual crop within the area that they occupy per m^3" (Garrity *et al.*, 2010).

Legume tree-based farming systems offer an important route to increase the availability of nitrogen while avoiding synthetic fertilisers. This has led to the use of the term *fertiliser tree* (Garrity *et al.*, 2010). Depending on tree species and soil status, fertiliser trees can fix from 5 to over 300 kg N ha^{-1} yr^{-1} (Rosenstock *et al.*, 2014). The use of *Gliricidia sepium* in improved fallows resulted in a 55 per cent increase in sorghum yield over two cropping seasons (Hall *et al.*, 2006). Sileshi *et al.* (2012) compared yields across three systems: maize-*Gliricidia* (the agroforestry cohort), conventional monoculture (with fertilisation) and regular practice (absent any external input). Yields in the agroforestry system were comparable to those achieved via synthetic fertilisation, but 42 per cent higher than non-fertilised fields. They were also more stable over time than yields achieved through synthetic fertilisation.

The presence of fertiliser trees delivers several other social, economic and environmental benefits. Increased availability and proximity of firewood has reduced the foraging time for women, and increased availability of fodder has increased their ability to keep livestock and therefore boosted their income. Environmental gains include increases to the water table, decreased soil erosion, significant carbon sequestration and increased soil nitrogen. Fertiliser trees are being adopted as a result of participatory trials and on-farm testing. As a result of direct stakeholder involvement, many innovations have come from farmers themselves.

In eastern Zambia, the average net profit was US$130 per ha when farmers cultivated maize without fertiliser; US$309 when it was grown in rotation with *Sesbania*; and US$327 when it was intercropped with *Gliricidia*. The benefit–cost ratio for farmers showed that integrated trees were more favourable than for those who used subsidised mineral fertiliser (Ajayi *et al.*, 2009). Adopting agroforestry practices has helped smallholder farmers in East and Southern Africa to overcome a major barrier to the adoption of conservation agriculture: the lack of crop residues to maintain a constant soil cover. As most African smallholders also raise livestock, they often use crop residue biomass as animal fodder. Now, with trees growing on their farms, there is enough biomass to meet different needs. The trees also provide fuel for rural households: in Zambia, farmers were able to gather 15 tonnes of fuelwood per hectare after the second year of fallow with *Sesbania* and 21 tonnes after the third year (Garrity *et al.*, 2010).

Legume-based agroforestry systems could result in increased N$_2$O emissions (Hall *et al.*, 2006), depending on species, soil type, climatic conditions and management practices. Based on a review of evidence from agroforestry systems in tropical regions and improved fallows in sub-Saharan Africa, Rosenstock *et al.*

(2014) concluded: "legume-based agroforestry is unlikely to contribute an additional threat to increasing atmospheric N_2O concentrations, by comparison to the alternative (e.g. mineral fertilizers)." Legume-based tree systems also sequester and accumulate carbon, and may also affect methane exchanges. Overall, there is a need for further research on the precise impacts of changes to C and N cycling, and an exploration of other environmental trade-offs such as increased leaching of N into local water supplies.

Patch intensification

The use of small plots of land to cultivate crops or rear fish, poultry or small livestock near places of human settlement represents the oldest form of agriculture (Niñez, 1984), and are amongst the most diverse and productive (per unit area) cultivation systems in the world (Conway, 1997). They provide several significant nutritional, financial and ecosystem benefits including pollination; gene-flow between plants inside and outside the garden; control of soil erosion and improvements to soil fertility; improved urban air quality; carbon sequestration and temperature control through the creation of microclimates. Patches contribute directly to household food and nutritional security, by increasing the availability, accessibility and utilisation of nutrient-dense foods (Galhena *et al.*, 2013), including wild edible species and traditional varieties no longer cultivated on commercial scale (Galluzzi *et al.*, 2010).

Tropical home gardens are typically multilayered environments with multiple trophic levels and management zones, and can include fruit trees, shaded coffee, residential, ornamentals with shade trees, multipurpose trees, herbaceous crops, ornamentals with vine-crop shade, grass, space for working and storage, and ornamentals. Such agroforestry home gardens can host over 300 plant species (Méndez *et al.*, 2001). In Peru, four kinds of small food production systems have been documented: fenced gardens near homes, plots planted as gardens near fields, field margins cropped with vegetables, and intercropping of the outer rows of staple fields with climbing vegetables (Niñez, 1984). In Java, *pekarangan* agroforestry gardens provide a safeguard against crop failure, and Tanzanian *chagga* gardens produce 125 kg beans, 275 bunches of banana and 280 kg of parchment coffee annually on plots of just over half a hectare, insuring against crop failure and supporting poultry and small livestock (Niñez, 1984).

Patches are also important in urban settlements. An estimated 800 million people practise some form of urban food production around the world, most on relatively small patches of land cultivated for subsistence (Lee-Smith, 2010). In Peru and Brazil, urban home gardens increased the availability of staple and non-staple foods to slum dwellers (Niñez, 1985; WinklerPrins, 2003; WinklerPrins and de Souza, 2005). Small patches are also important for household resilience during lean seasons, in conditions of political instability and turmoil, for marginalised households, in degraded or highly populated areas with few endowments of land and materials and in disaster, conflict and post-crisis situations. Examples include the use of gardens for food during the Tajik civil war (Rowe, 2009); the conflicts in Sri Lanka (Niñez, 1984); the use of relief and victory gardens in the USA and UK during the world wars, and the use of gardens to tide over political and economic crisis in Cuba (Pretty, 2002). Food growing on neglected patches of city land, along highways "often represent the only green spots in abandoned and neglected city parks" (Niñez, 1984).

In recognition of these benefits, home-gardening and the cultivation of small patches around fields has been promoted in development initiatives to improve food security and incomes. Across Africa, an important strategy has been the construction of raised beds to improve the retention of water and organic material (Pretty *et al.*, 2011a). In Kenya and Tanzania, the FarmAfrica project has encouraged the cultivation of African indigenous vegetables by 500 participating farmers who have used 50 per cent less fertiliser and 30 per cent less pesticide than conventionally grown vegetables (Muhanji *et al.*, 2011). In Bangladesh, the Homestead Food Production has involved 942,040 households (some 5 million beneficiaries) between 1988 and 2010 (Iannotti *et al.*, 2009). Through home-gardening and small-animal husbandry, the project has achieved notable success in increasing the production of fruit, vegetables and eggs relative to controls, and increased income from the sale of produce.

Raised beds, Kenya

Double-dug raised beds were first widely promoted in Kenya by Manor House and the Association for Better Land Husbandry. These involve investments in very small patches of the land, though some professional agriculturalists dismiss this as only gardening. Raised beds are combined with composts and animal manures to improve the soil. A considerable initial investment in labour is required, but the better water-holding capacity and higher organic matter means that they are able to sustain vegetable growth long into the dry season. Once the investment is made, little more has to be done for the next two to three years. Many vegetable and fruit crops can be cultivated, including kales, onions, tomatoes, cabbage, passion fruit, pigeon peas, spinach, peppers, green beans and soya. Self-help groups have found that their family food security has improved substantially. Families are now finding that by working more on their own farms rather than selling labour to others, they are getting greater returns. Children have been beneficiaries, as their health has improved through increased vegetable consumption and longer periods of available food. In 26 communities, 75 per cent of households are now free from hunger during the year, and the proportion of households having to buy vegetables has fallen from eight to just one out of ten. Manor House has trained 70,000 farmers in constructing and cultivating double-dug beds.

Many agriculturalists had been sceptical about these methods, saying they needed too much labour, were too traditional and had no impact on the rest of the farm. Yet the main benefits are for women. In Kakamega, Joyce Odari built 12 raised beds on her farm. They were so productive that she employed four young men from the village. She said: "if you could do your whole farm with organic approaches, then I'd be a millionaire. The money now comes looking for me". She was also aware of the wider benefits: "my aim is to conserve the forest, because the forest gives us rain. When we work our farms, we don't need to go to the forest. This farming will protect me and my community, as people now know they can feed themselves". Another example is Susan Wekesa, who said she had "moved my household from misery to normal rich life. My *shamba* is producing a surplus, which I sell for income ... I can now face the future proudly". She had a tiny amount of land: just over a tenth of a hectare. Once again, the spin-off benefits were substantial, as giving women the means to improve their food production meant that food got into the mouths of children. They suffered fewer months of hunger, and so were less likely to miss school.

Integrated farming on small plots, China

In the shadow of the Great Wall, Bei Guan village lies in the rolling hills and plains of Yanqing County, and is the site for one of the experiments in the integration of sustainable agriculture and renewable energy production. It was selected by the Ministry of Agriculture as an ecological demonstration village in one of 150 counties across the country for implementation of integrated farming systems.

Farmers have made the transition from monocultural maize cultivation to integrated vegetable, pig and poultry production. Each of the 350 households has a small plot of two mu (a seventh of a hectare), a pen for livestock and a biogas digester. Ten vegetables are grown and sold directly into Beijing markets. Green wastes are fed to the animals, their wastes channelled to the digester, which in turn produces methane gas for cooking, lighting and heating; the solids from the digester then being returned to the soil. Each farmer also uses plastic sheeting to create greenhouses from the end of August to May, thus extending production through winter when temperatures regularly fall to −30 degrees centigrade.

The advantages for local people and the environment are substantial: more income from the vegetables, better and more diverse food, reduced costs for fertilisers, reduced workload for women, and better living conditions in the house and kitchen. In Bei Guan, there is also a straw gasification plant that uses only maize husks to produce gas to supplement household production. Instead of burning husks in inefficient stoves, requiring 500 baskets per day for the whole village, now just 20 are burnt per day in the plant. The village head, Lei Zheng Kuan, noted "these have saved us a lot of time. Before, women had to rush back from the fields to collect wood or husks, and if it had been raining, the whole house would be full of smoke. Now it is so clean and easy".

The benefits of these systems are far reaching. The Ministry promotes a variety of integrated models across the country, involving mixtures of biogas digesters, fruit and vegetable gardens, underground water tanks, solar greenhouses, solar stoves and heaters, and integrated pig and poultry systems. These are fitted to local conditions. Whole integrated systems are now being developed across many regions of China, and altogether 50 million households now have digesters. As the systems of waste digestion and energy production are substituting for fuelwood, coal or inefficient crop-residue burning, the benefits for the natural environment are substantial – each digester saves the equivalent of one and a half tonnes of wood per year, or three to five mu of forest. Each year, these biogas digesters are effectively preventing carbon from being emitted to the atmosphere.

Small-scale and integrated aquaculture

Aquaculture, especially when integrated within crop cultivation, is also a form of patch management. In the last two decades, dramatic growth in aquaculture production has boosted average consumption of fish and fishery products at the global level. The shift towards relatively greater consumption of farmed species compared with wild fish reached a milestone in 2014, when the farmed aquaculture sector's contribution to the supply of fish for human consumption surpassed that of wild-caught fish for the first time (FAO, 2016e). Inland aquaculture produces 42 Mt annually, marine aquaculture a further 25 Mt. Integrated aquaculture–agriculture (IAA) has been developed and promoted in Malawi and Cameroon (Brummett and Jamu, 2011). The WorldFish Center, supported by donors and partnered by international and national research and rural development agencies as well as local farmers, have helped in the expansion of IAA from the 1990s. Where successful, IAA has been found to contribute to "a 40% improvement in farming system resilience (defined by the ability to maintain positive cash flows through drought years), a 50% reduction in nitrogen loss and improved nitrogen-use efficiency". Agricultural productivity has increased, as has per capita farm income and per capita consumption of fish.

In Nigeria, the rise of peri-urban aquaculture brings into focus several firsts in African aquaculture development (Miller and Atanda, 2011). Aquaculture has progressed from a subsistence-focused activity to one connected to a growing market: Miller and Atanda report "a remarkable 20% increase in growth per year over six years, with high growth in small-to-medium-scale enterprises and a number of large-scale intensively managed fish farms". Aquaculture here is primarily focused on the African catfish (*C. gariepinus*), and is characterised by a highly developed production chain of suppliers, processors and marketers, supported by private-sector technical staff, a growing competence with public extension services, government support via livelihood programmes and credit support from lenders. Nigerian catfish aquaculture is estimated to provide some 17 per cent of the country's total domestic fish production. Producers have been able to diversify into further animal husbandry, keeping cows, pigs, goats and sheep. Aquaculture also supports crop cultivation, via the supply of irrigation and fertilisation from fish ponds.

Rice–fish systems in Asia

A field of rice in standing water is more than a crop: it is an ecosystem teeming with life, including duck, fish, frog, shrimp, snail and dozens of other aquatic species. For thousands of years, rice farmers have harvested that wealth of aquatic biodiversity to provide their households with a wide variety of energy- and nutrient-rich foods. The traditional rice–fish agroecosystem supplied micronutrients, proteins and essential fatty acids that are especially important in the diets of pregnant women and young children (Halwart, 2013). Farmers benefit from lower costs, higher yields and improved household nutrition. Combining rice and aquaculture also makes more efficient use of water. However, rice–fish farming requires about 26 per cent more water than rice monoculture and therefore these systems are not recommended where water supplies are limited. Nevertheless, an estimated 90 per cent of the world's rice is planted in environments that are suitable for the culture of fish and other aquatic organisms.

During the 1960s and 1970s, traditional farming systems that combined rice production with aquaculture began to disappear, as policies favouring the cultivation of modern high-yielding rice varieties – and a corresponding increase in the use of agrochemicals – transformed Asian agriculture. As the social and environmental consequences have become more apparent, there is renewed interest in raising fish in rice fields (FAO, 2014b; Ketelaar et al., 2018). There are two main rice–fish production systems. The most common is concurrent culture, where fish and rice are raised in the same field at the same time; rotational culture, where the rice and fish are produced at different times, is less common. Both modern short-stem and traditional long-stem rice varieties can be cultivated, as can almost all the important freshwater aquaculture fish species and several crustacean species (FAO, 2004, 2014b).

In China, rice farmers raise fish in trenches up to 100 cm wide and 80 cm deep, which are dug around and across the paddy field and can occupy some 20 per cent of the paddy area. Bamboo screens or nets prevent fish from escaping. While fish in traditional rice–fish systems feed on weeds and by-products of crop processing, more intensive production usually requires commercial feed. With good management, a one-hectare paddy field can yield 225–750 kg of finfish or crustaceans a year, while sustaining rice yields of 7.5–9 tonnes (FAO, 2007). In Vietnam, recently developed integrated rice–fish systems yield $7,700 of gross income per hectare, compared with $1,900 for rice alone (Ketelaar et al., 2018). On average these are producing 4.9 tonnes of fish and other aquatic organisms per hectare annually.

The combination of different plant and animal species makes rice–fish systems productive and nutritionally rich. These have been called co-culture technologies (Hu et al., 2016), where fish can include carp, crab, crayfish, even turtle. Equally important are the interactions among plant and animal species, which improve the sustainability of production. Studies in China found that the presence of rice stem-borers was around 50 per cent less in rice–fish fields. A single common carp can consume up to 1,000 juvenile golden apple snails every day; the grass carp feeds on a fungus that causes sheath and culm blight. Weed control is generally easier in rice–fish systems because the water levels are higher than in rice-only fields. Fish can also be more effective at weed control than herbicides or manual weeding. By using fish for integrated pest management, rice–fish systems achieve yields comparable to, or even higher than, rice monoculture, while using up to 68 per cent less pesticide. This safeguards water quality as well as biodiversity.

The interactions among plant and animal species in rice–fish fields also improve soil fertility. The nutrients in fish feed are recycled back into fields through excreta and made immediately available to the rice crop. Reports from China, Indonesia and the Philippines indicate that rice–fish farmers' spending on fertiliser is lower. Cultivating fish reduces the area available for planting rice. However, higher rice yields, income from fish sales and savings on fertiliser and pesticide lead to returns higher than those of rice monoculture. Profit margins may be more than 400 per cent higher for rice farmers culturing high-value aquatic species.

Raising fish in rice fields also has community health benefits. Fish feed on the larval vectors of infectious diseases, particularly mosquitoes that carry malaria. Field surveys in China found that the density of mosquito larvae in rice–fish fields was only a third of that found in rice monocultures. In one area of Indonesia, the prevalence of malaria fell from 16.5 per cent to almost zero after fish production was integrated into rice fields (FAO, 2004). In China, aquaculture in rice fields has increased steadily over the last two decades, and production reached 1.2 Mt of fish and other aquatic animals in 2010. New opportunities for diversifying production are opening up in Indonesia, where the tutut snail, a traditional item in rural diets, is becoming a sought-after health food for urban consumers. The resurgence in rice–fish farming is being actively promoted by the Government of Indonesia, which recently launched a 'one-million hectare rice–fish programme'. While there is compelling evidence of the social, economic and environmental benefits of aquaculture in rice farming systems, its rate of adoption remains low outside of China. Elsewhere in Asia, the area under rice–fish production is only slightly more than 1 per cent of the total irrigated rice area. Interestingly, the rice–fish farming area is proportionally largest outside Asia, in Madagascar, at nearly 12 per cent.

Yet, there is much scope for wider uptake of these systems. Improved awareness of its benefits is required. Where smallholders can access low-cost pesticides, these may lock-in rice monocultures. Farmers may also have limited access to credit for investment in fish production. Overcoming these barriers is difficult because multi-sectoral policymaking is involved. Rice–fish farming needs to be championed by agricultural policymakers and agronomists who recognise the benefits of integrating aquaculture and rice, and can deliver that message to rice-growing communities. Just as agricultural development strategies once promoted large-scale rice monoculture, they can now help to realise the potential of intensive, but sustainable, rice–fish production systems.

Systems of rice and crop intensification (SRI and SCI)

A radical approach to crop intensification was begun two decades ago with revisions to the agronomy and agroecology of rice cultivation. This was called the System of Rice Intensification (SRI), and was later spread to a range of other crops (System of Crop Intensification). SCI is an agricultural production strategy that seeks to increase and optimise the benefits that can be derived from making better use of available resources: soil, water, seeds, nutrients, solar radiation and air (Adhikari *et al.*, 2018).

SCI principles and practices build upon the productive potentials that derive from plants having larger, more efficient, longer-lived root systems and from their symbiotic relationships with a more abundant, diverse and active soil biota. The main elements comprise:

- Starting with high quality seeds or seedlings, well selected and carefully handled, to establish plants that have vigorous early growth, particularly of their root systems.
- Providing optimally wide spacing of plants to minimise competition between plants for available nutrients, water, air and sunlight. This enables each plant to attain close to its maximum genetic potential.
- Keeping the topsoil around the plants well aerated through appropriate implements or tools so that soil systems can absorb and circulate both air and water. This is usually done as part of weeding operations, and can stimulate beneficial soil organisms, from earthworms to microbes, at the same time reducing weed competition.
- If irrigation facilities are available, these should be used but sparingly, keeping the soil from becoming waterlogged and thus hypoxic. A combination of air and water in the soil is critical for plants' growth and health, sustaining both better root systems and a larger soil biota.
- Amending the soil with organic matter, as much as possible, to enhance its fertility and structure and to support the soil biota. Soil with high organic content can retain and provide water in the root zone on a more continuous basis, reducing crops' need for irrigation water.
- Reducing reliance on inorganic fertilisers and pesticides and, if possible, eliminating them altogether.

The merit of this kind of agroecological approach for achieving more productive phenotypes from given genotypes of rice has been validated through a number of well-designed agronomic studies (e.g. Lin *et al.,* 2009; Zhao *et al.*, 2009; Thakur *et al.*, 2010a, 2010b, 2011) as well as for wheat (Dhar *et al.*, 2016). Adaptation to suit local contexts and farmers' preferences is encouraged, and so participatory models of development and dissemination have been important for the spread of SRI approaches, resulting in adaptations that enable farmers to increase yields, lower resource-use and buffer against risk in ways that suit them (Krupnik *et al.*, 2012).

Many of these agroecologically based processes are still not completely understood, though it is now clear that the interactions between crop plants and the soil biota with respect to water and nutrient uptake will be enhanced by having individual plants with expanded root systems and a more active and diverse soil biota (Lin *et al.*, 2009; Anas *et al.*, 2011; Barison and Uphoff, 2011; Thakur *et al.*, 2013).

These positive interactions are complemented by the beneficial effects of symbiotic bacterial and fungal endophytes (Thakur et al., 2016).

System of rice intensification

Traditionally, rice has been cultivated in Asia mostly by first puddling fields, which means flooded then ploughed, to create a soft, muddy soil layer often overlying a hard pan, which restricts downward flow and loss of water. Rice seedlings 20–60 days old are then transplanted to the field, in clumps of up to eight plants, randomly distributed or in narrowly spaced rows. Until the crop matures, the paddy is continuously flooded with 5–15 cm of water to suppress weeds. This system has enabled the cultivation of rice for millennia at low, but relatively stable, yields. When the Green Revolution introduced high-yielding varieties, mineral fertiliser and chemical pest control, per hectare productivity in many Asian rice fields doubled over 20 years (Papademetriou, 2000).

A set of crop, soil and water management practices known as the System of Rice Intensification takes a strikingly different approach. Seedlings 8–15 days old are transplanted singly, often in grid patterns with wide spacing of 25 x 25 cm per plant. To promote moist, but aerated, soil conditions, intermittent irrigation is followed by dry periods of 3–6 days. Weeding is done at regular intervals, and compost, farmyard manure and green manure are preferred to mineral fertiliser. Once the plants flower, the field is kept under a thin layer of water until 20 days before the harvest.

SRI emerged from on-farm experimentation in Madagascar whereby existing norms of paddy rice were radically amended: reduced planting density, improvement of soil with organic matter, reduced application of water and very early transplantation of young plants. The adoption of these four general principles have been shown to lead to considerable increases in yields with reduced inputs of water and other external inputs (Uphoff, 1999, 2003; Stoop et al., 2002). Since its inception in the 1980s, SRI principles have been adapted and applied to a variety of other crops, including wheat, sugarcane, tef, finger millet, various pulses and turmeric, all again emphasising changes in resource use and application combined with crop planting design.

SRI remains, however, the subject of some controversy, largely because yield increases can only partly be explained. Important questions remain related to the agronomy of SRI and to claims of large yields (Dobermann, 2004; Sheehy et al., 2004, 2005; Stoop 2011). There has also been debate as to whether yields under SRI compare favourably with best-recommended practice (McDonald et al., 2006, 2008; Uphoff et al., 2008).

In India, experimental comparisons between SRI and conventional recommended practice showed 42 per cent higher yields under SRI as a result of changes in plant physiological processes and characteristics including longer panicles, more grain per panicle, a higher proportion of grain-filling, deeper and better distributed root system, and more and larger leaves (Thakur et al., 2010a, 2010b). Another Indian study compared SRI and conventional cultivation across 13 rice-growing states and showed between 12–54 per cent higher yields with SRI, combined with improved water-use efficiency (Palanisami *et al.*, 2013).

As with other agroecological approaches, SRI practices are flexible. In Mali, Styger *et al.* (2011) reported increased yields without the addition of expensive inputs in the Timbuktu region, where traditional rice cultivation has depended on the annual flooding of the Niger River. Africare has worked with farmers to adapt SRI principles to local conditions and evaluate the potential of SRI to increase rice yields and thus food security in the region: SRI plots yielded 66 per cent more than control plots and 87 per cent more than surrounding rice fields. SRI plots used substantially fewer seeds per hectare (85–90 per cent less than conventional plots), 30 per cent less inorganic fertiliser and 10 per cent less irrigation water. However, a meta-analysis by Turmel *et al.* (2011) of data from 72 studies comparing SRI and conventional practice found that SRI outperformed conventional practice on weathered and infertile soils, but showed no yield advantage on more favourable soils.

Other systems of rice cultivation that save water and labour include direct seeding, whereby rice is sown and sprouted directly into the field rather than raising and transplanting seedlings. Direct seeding has long been practised traditionally, and is undergoing a resurgence in response to water scarcity and labour shortages in Asia (Rao *et al.*, 2007). In Sri Lanka, some 95 per cent of all rice grown is direct-seeded under either wet- or dry-seeding (Weerakoon *et al.*, 2011). Weed infestations and relatively low or variable yields (compared with those under traditional transplantation) pose problems in direct-seeded systems. Reviews suggest the need for the development and dissemination of improved cultivars (including herbicide-resistant and early maturing varieties), improved nutrient management and the provision of high quality herbicides (Farooq *et al.*, 2011).

The governments of Cambodia, China, India, Indonesia and Vietnam have endorsed SRI methods in their national food security programmes, and millions of rice farmers have adopted SRI practices in their own fields. In Vietnam, the Ministry of Agriculture and Rural Development launched farmer training in SRI in 2003, with FAO assistance. After an evaluation found average yield increases of 11 per cent, along with major reductions in fertiliser, pesticide and water use and a 50 per cent increase in farmer incomes, the ministry scaled up the initiative to the national level (Africare *et al.*, 2010).

By 2011, 1 million Vietnamese rice farmers were applying SRI practices. Farmers' net profits increased by an average of US$110 per hectare, owing to a 40 per cent reduction in production costs (Dũng *et al.*, 2011). Farmers who were trained in site-specific nutrient management benefited from additional annual net income of up to US$78 per hectare (Nga *et al.*, 2010).

Uptake of SRI has been similarly rapid and widespread in India. Tests of a package of SRI practices in Tamil Nadu state in 2003 resulted in grain yield increases of 1.5 tonnes per ha. With state government support, the system has since been adopted on an estimated 600,000 ha of rice lands, with yields of up to 9 tonnes per hectare (Africare *et al.*, 2010). The government of Andhra Pradesh state in 2005 launched a programme to establish SRI demonstration plots in every village in the state. An estimated 600,000 Indian farmers grow their rice using all or most of the recommended SRI practices (Africare *et al.*, 2010).

From many evaluations of rice, it is now clear that yields can be boosted by 25–50 per cent by agroecological management, and often the increases are measured at a doubling or more. These effects are quite explainable in scientific terms (Thakur *et al.*, 2016). Crops that have better developed root systems, for example, are less vulnerable to drought and to lodging (being knocked down by wind or rain). They are also generally more resistant to attacks and losses from pests and diseases. In addition to enabling crops to resist the stresses of climate change, there are net reductions in greenhouse gas emissions (Thakur and Uphoff, 2017; Uphoff, 2015).

Wider applications of system of crop intensification

It is now being found that the components of SRI can be extended beyond rice. Farmers in a dozen countries have applied the concepts and practices of SRI with appropriate modifications to a variety of crops. Although most such applications began less than a decade ago, they have burgeoned. A summary of impacts and extent is shown in Table 5.2. These developments are widespread: the majority of innovation is occurring in India, but also in Afghanistan, Cambodia, Nepal and Pakistan in Asia; in Ethiopia, Kenya, Malawi, Mali and Sierra Leone in Africa; and in Cuba in Latin America, with SCI getting started also in the USA.

Initially it was thought that methods which succeeded with rice would apply only to other crops in the botanical family of grasses, such as wheat, barley, millet, tef, sugarcane. These monocotyledons have multiple, roughly parallel tillers, and thick, bushy root systems, rather than growing with dominant main stems and main (tap) roots from which branching canopies and root systems emerge. However, a number of dicotyledonous crop plants such as mustard, legumes, green leafy vegetable and some spices have also responded positively to SCI practices. Farmers have seen improvements in yield, profitability and resilience when they have extrapolated SRI practices to widely varying crop types, either on their own or with encouragement from civil society, government or university partners.

Integrating system components

We now reflect on some examples where a variety of system components have been integrated to address soil conservation at watershed levels, to integrate legumes and green manures into cereal-based agroecosystems, and to maximise the outcomes for livestock in maize and legume systems.

A recent meta-analysis of intercropping in African systems found that it on average increased crop yields by 23 per cent, and brought gross income increases of $170 per hectare (Himmelstein *et al.*, 2017). In Asia, intercropping rice and water spinach also had a dramatic effect on productivity and income. Growing the wet

TABLE 5.2 Effects of SCI practices on yield across eight crops in seven countries

Crop	Country (state, province or district)	Comparison yield (t/ha)	SCI yield (t/ha)	Extent and/or sources of improvements
Finger millet	India (Karnataka)	1.25–2.5 (3.75 max)	4.5–5.0 (6.25 max)	Indigenous farmer-developed system of cultivation (*Guli Ragi*)
	Ethiopia (Tigray)	1.3	4.0–5.0	SCI methods developed before any knowledge of SCI
	India (Uttarakhand)	1.5–1.8	2.4	Evidence of climate resilience
	India (Odisha)	1.0–1.1	2.1–2.25	Evidence of climate resilience; 2013: 143 farmers; 2016: 2,259
Wheat	India (Bihar)	2.0	4.6	2008/09: 278 farmers; 2015/16: ~500,00 farmers (300,000 ha)
		2.25	3.87	86% increase in income/ha
	Nepal (Khailali and Dadeldhura)	3.4	6.5	Replicated trials, average for upland and lowland yields
	Afghanistan	3.0	4.2	Farmer results under national FAO programme
	Mali (Timbuktu)	1.0–2.0	3.0–5.0	Farmer trials under Africare
	India (IARI research)	5.42	7.44	Two years of experimental results at Indian Agricultural Research Institute, New Delhi
Maize	India (Him. Pradesh)	2.8	3.5	On-farm trials
	India (Assam)	3.75–4.0	6.0–7.5	On-farm trials
Sugarcane	India (Telengana)	80	99.5	On-farm trials
	India (Odisha)	60–70	119	On-farm trials
	India (Maharashtra)	70	96	On-farm trials
	India (Uttar Pradesh)	61 [59]	68 [71]	On-farm trials [ratoon harvest]
	Kenya (Kakamega)	70	90–100	On-farm trials
	Cuba	60–75	85–100	Modified SSI on-farm trials
Tef	Ethiopia	1.6	2.8 (TIRR)[1] 3.0–5.0 (STI)[2]	Estimated TIRR area in 2016 was 1.1 million ha

(*Continued*)

TABLE 5.2 (Continued)

Crop	Country (state, province or district)	Comparison yield (t/ha)	SCI yield (t/ha)	Extent and/or sources of improvements
Mustard	India (Bihar)	1.0	3.0 4.0 (full use)	On-farm production
	India (Mad. Pradesh)	1.2	2.73 (1.8–3.3)	On-farm trials
Pulses	India (HP/UKD/MP)	46% average increase across seven pulses		
	India (Bihar)	56% increase across different pulses (15,590 households)		
Vegetables	India (Bihar)	20% average increase in yield; 47% increase in net income/ha		

Source: From Adhikari *et al.* (2018).

Notes
1. TIRR = Teff, Improved seed (Quncho), Reduced seed rate, and Row planting.
2. STI = System of Teff Intensification.

rice with water spinach reduced rice blast and sheath blight disease, increased rice tiller numbers, and raised net photosynthesis. Net income increased by fivefold (Liang *et al.*, 2016).

In China, Thailand and Vietnam, nectar-producing plants were introduced around rice fields, and it has been found that this landscape diversification promotes ecological intensification (Gurr *et al.*, 2016). Over four years, pest incidence decreased, pesticide use was lowered by 70 per cent and rice grain yields increased by 5 per cent to 6.6 tonnes per hectare. The intervention of border vegetation had increased the populations of natural enemies.

Greening the drylands of West Africa

The West African Sahel has been the site of remarkable improvements, particularly within Burkina Faso and Niger, as a result of soil and water conservation across some 4 Mha and the planting of approximately 120 million trees through agroforestry. In Niger, some 300,000 ha of degraded land has been rehabilitated, and much of the landscape has been transformed as compared with its status in the 1970s. The adoption of tree planting has seen significant increases since the decentralisation of power and ownership rights. In the 1980s, farmers thought that trees belonged to the state. The follow-on benefits from agroforestry are significant, and include rejuvenated aquifers, increased availability of fodder and fuelwood and a concomitant decrease in the amount of time spent by women in collecting firewood. Women's incomes have also increased as a result of these initiatives, as the increased availability of fodder allows them to own livestock. Soil conservation structures, together with the planting of trees and the application of organic fertiliser, have been instrumental in stabilising and increasing the yields of staple crops in Burkina Faso and are thus a vital support for food security in the country.

Soil conservation initiatives, in combination with water conservation and the addition of new system components such as trees and livestock, can improve the productivity of staples and increase farmers' incomes. Soil conservation is a collective action challenge. As it usually takes place across whole landscapes, it requires the cooperation and active participation of several landowners.

A Foresight case study (Sawadogo, 2011) describes how soil and water conservation techniques, in conjunction with tree planting, have rehabilitated degraded land in Burkina Faso and thus contributed to food security and poverty alleviation. Traditional techniques include the construction of zaï, rock bunds, stone lines and half-moons. Zaïs are traditional constructions used to increase soil moisture and organic content. Farmers place organic matter, such as composted manure, into circular holes 20 cm across and 15 cm deep. These create a micro-environment that allows crops such as sorghum and millet to flourish despite poor soils and dry conditions. Traditional zaï constructions have been improved by individual innovators and technical experts from Oxfam UK's agroforestry project and subsequently spread. The technology is currently used extensively across north-western Burkina Faso, covering between 30,000 and 60,000 ha. Tests of the impact on soil fertility and yields show that zaï can produce significant increases in soil moisture and organic content; under certain conditions, crop yields have been shown to double or even quadruple.

Rock bunds, another traditional soil conservation measure, were also developed and spread across Burkina Faso via training programmes. As with zaï, rock bunds have a significant impact on yields. In regions where soils were treated with rock bunds and compost manure, yields have increased by 77–130 per cent. Half-moons were first introduced in the late 1950s. These consist of semicircular bunds, within which dryland crops such as sorghum and millet can be planted. The spread of training programmes such as farmer field schools has ensured that the approach is now widely adopted in Burkina Faso.

Intercropping with the fertiliser tree, *Faidherbia albida,* presents one of the few examples of intercropping being practised at scale (Montpellier Panel, 2013), resulting in what has been called a green wall for the Sahel (Reij and Smalling, 2008). *Faidherbia* is a nitrogen-fixing acacia indigenous to Africa. The tree is particularly suited to intercropping with maize, as it sheds its leaves during the monsoon season when maize is sown. This prevents it from competing with maize seedlings for light and nutrients, while falling leaves provide nutrients. Intercropping with *Faidherbia* can result in cereal yields of 3 t/ha without the application of additional fertiliser, while contributing to significant carbon sequestration, weed suppression, increased water filtration and increased adaptability to serious droughts (Garrity *et al.*, 2010). In Zambia and Malawi, the planting of *F. albida* within maize fields is practised over some 500,000 hectares. Niger has experienced the most remarkable re-greening as a result of farmer-managed regeneration of trees in fields.

As a result of the relaxation of restrictive policies prohibiting farmers from managing the trees on their own lands, agricultural landscapes in Niger now contain significant densities of *Faidherbia* (Garrity *et al.*, 2010). It has been estimated that re-greening has resulted in a per year increase of 500,000 additional tonnes of food (Reij *et al.*, 2009). Maize, sorghum, millet, groundnuts and cotton have all shown increased yields, even without additional fertiliser application, as a result of *Faidherbia*. In Burkina Faso, *Crombretum glutinosum* and *Piliostigma reticulatum* are included to generate additional biomass and yield fodder and increased yields.

Slash-and-mulch green manures, Honduras and Guatemala

Since the collapse of the Mayan civilisation, indigenous people have farmed with slash-and-burn methods. Fields are cleared in the forest, cropped for a couple of years, and then abandoned as families move on to new sites. Over time, as the population has increased, and as others came to log the forests, so farmers have had to reduce fallow periods, thus returning to former fields too soon for natural soil fertility to have been restored. Both farming and forest come under pressure: yields remain low or fall, the forest steadily disappears.

There have been several approaches to developing intensive settled agroecosystems that are more productive (Bunch, 2018). First, the velvetbean (*Mucuna pruriens*) was introduced in maize systems in Honduras and Guatamala, particularly by the non-governmental organisations (NGOs) World Neighbors, Cosecha and Centro Maya. They found that cultivation with maize substantially increased cereal yields. Mucuna is grown as a soil improver. It can fix 150 kg of nitrogen per hectare and produce 50–100 tonnes of biomass annually. This plant material falls on the soil as a green manure, suppressing weeds and helping to build the soil. Build the health of the soil, and farmers no longer need to burn trees to create new fields. Such improvements to soil health change the way farmers think. They see the benefit of staying in the same place, and of investing in the same fields for themselves and their children. At the Usumacinta River border of Guatemala with Mexico, Centro Maya work with the Cooperativa La Felicidad (Happiness Cooperative), where 250 farmers grow mucuna in maize, and have begun a journey across a cognitive frontier towards settled and sustainable agriculture. Mucuna is called the bean manure (*fritjol abono*): farmers say the bean manure destroys the weeds, the beans simply kill them, and all the crops flourish much more. It is technically easy: improve soils through low-cost methods, and rainforests can be saved.

Similarly, farmers in the village of Quezungual in Honduras developed a low-cost, resource-conserving system for growing their crops (Ayarza and Welchez, 2004). Instead of clearing forest and burning vegetation, they adopted a slash-and-mulch approach. They begin by broadcasting sorghum or beans in an area of well-developed, naturally regenerated secondary forest. After planting, they selectively cut and prune the trees and shrubs, and spread the leaves and small branches on the soil surface to create a layer of mulch. High-value timber, fruit and fuelwood trees are left to grow (Ayarza and Welchez, 2004; CIAT, 2009). Once the sorghum and beans have been harvested, maize is planted. Farmers continue to slash and prune trees to ensure the crops get sufficient sunlight, while leaves, branches and crop residues are used to maintain a semi-permanent soil cover. The soil is not tilled, and fertiliser is applied only when needed (CIAT, 2009).

In the early 1990s, FAO began working closely with local farmers and farmers' groups to develop and disseminate those practices, which have become known as the Quesungual Slash-and-Mulch Agroforestry System (QSMAS). The system has since been adopted by more than 10,000 low-income farmers in south-west Honduras. Using QSMAS, farmers improved the crop yields normally produced by slash-and-burn methods by more than 100 per cent: maize from 1.2 t to 2.5 t per ha, and beans from 325 to 800 kg per ha (Ayarza and Welchez, 2004). Increased productivity has improved food security and has allowed farmers to set aside space in their fields to explore different options for producing food. Almost half of the farmers who have adopted QSMAS use some part of their land, and their additional income, to diversify production, primarily with home gardens and livestock (CIAT, 2009).

Farmers have embraced the system because it is founded on familiar, indigenous farming practices, is more productive and profitable than slash-and-burn agriculture, and delivers many other benefits. By retaining soil moisture and preventing erosion, QSMAS has made farms more resilient to extreme weather events, such as a drought and hurricane. The system also reduces the time required to prepare the land and control weeds, an important consideration in an area where labour scarcity is a major constraint to improving farm productivity. Rural communities also benefit from improved water quality, as well as more water availability during the November to April dry season. The trees retained on QSMAS farms meet around 40 per cent of households' fuelwood needs (CIAT, 2009). The success is also due to the fact that local communities and extension workers were encouraged to share ideas and learn from each other. Thanks to that participatory process, the impact has reached beyond the farmers' fields. Once they became more aware of the problems created by deforestation, community institutions banned the use of slash-and-burn (CIAT, 2009).

Maize–grass–livestock, Brazil

Livestock production is particularly important in many smallholder farming systems of Latin America. However, output per animal unit in tropical areas is far below that achieved in temperate regions. A major constraint is the quantity and quality of forage, a key feed source in ruminant systems. Overgrazing, farming practices that deplete soil nutrients and a lack of forage species that are better adapted to biotic and abiotic stresses – all contribute to low productivity. Improving pasture forage quality and productivity would help to boost production of meat and milk (Rao *et al.*, 2014).

Many livestock farmers in Latin America have adopted a sustainable livestock production system that integrates forages with cereals. A key component of the system is *Brachiaria*, a grass native to sub-Saharan Africa, which grows well in poor soils, withstands heavy grazing and is relatively free from pests and diseases. Thanks to its strong, abundant root system, *Brachiaria* is also efficient in restoring soil structure and preventing soil compaction.

The productivity of animals feeding on *Brachiaria* pastures is 5–10 times higher those feeding on native savannah vegetation, making a significant contribution to farmers' incomes. In Brazil, where about 100 Mha are planted with *Brachiaria*, annual economic benefits have been put at US$4 billion; in Colombia, they are thought to exceed US$1 billion (CIAT, 2013). In Central America *Brachiaria* pastures generate an additional value of about US$1 billion a year, with 80 per cent of the gains accruing to the beef sector and 20 per cent to the milk industry (Holmann *et al.*, 2004). While the adaptation of *Brachiaria* to low-fertility soils has led to its use for extensive, low-input pastures, it is also suitable for intensively managed pastures (Rao *et al.*, 2014). *Brachiaria* also has the ability to convert residual soil phosphorus into organic, readily available forms for a subsequent maize crop (Resende *et al.*, 2007).

The rotation of annual crops with grazed pasture is not commonly practised, but it is increasing in the Cerrados region of Brazil, where beef cattle are a major source of income for many farmers. Years of poor herd management, overgrazing and lack of adequate soil nutrient replacement have led to declining productivity and reduced profitability in traditional livestock production systems (Rao *et al.*, 2014; Klink and Moreira, 2002).

Where natural ecosystems have been replaced by intensive soybean monoculture, serious degradation has occurred. Much of the soil is compacted and susceptible to erosion from heavy rainfall. Under these conditions, traditional techniques of soil erosion control, such as contour planting, have proven to be inadequate (Pacheco *et al.*, 2013). In response, many farmers have adopted zero-tillage systems. It has been estimated that around 50 per cent of the total cropped area in Brazil is under direct-seeded mulch-based cropping (DMC) systems, which usually grow three crops a year, all under continuous direct seeding. In the Cerrado alone, more than 4 Mha are cultivated using diversified DMC systems, which have replaced inefficient, tillage-based soybean monoculture. A typical sequence is maize (or rice), followed by another cereal, such as millet or sorghum, or the grass *Eleusine*, intercropped with a forage species such as *Brachiaria* (Kluthcouski and Pacheco-Yokoyama 2006; Scopel *et al.*, 2004). The forages function as powerful 'nutrient pumps', producing large amounts of biomass in the dry season which can be grazed or used as green manure. The combination of maize plus *Brachiaria* at the end of the rainy season taps soil water from levels deeper than 2 metres, and promotes active photosynthesis later during the dry season. It results in vigorous vegetative re-growth after the first rains of the following season, or after rain during the dry season, thus ensuring permanent soil cover.

Because *Brachiaria* provides excellent forage for cattle, farmers can then choose to convert the area immediately into pasture, or keep it in grain production for another year. Such systems are found under irrigation or in wetter regions with frequent, heavy rains, which recharge deep water reserves. In the best DMC systems, total annual dry matter production, above and below the soil, averages around 30 tonne per ha, compared with 4–8 tonne found under monocropping. To reduce intercrop competition, novel intercropping systems have been developed. In the Santa Fé system for maize and *Brachiaria*, developed in Brazil (Kluthcouski *et al.*, 2000), the grass is made to germinate after the maize crop, either by delaying its planting or by planting it deeper. The young *Brachiaria* plants are shaded by the maize, and provide little competition for the cereal. At maize harvest, however, shading is reduced and the established pasture grows quickly over the maize residue. This tight integration between forage and grain crops leads to a better use of the total farm area and a more intensive use of the pastures, with less pasture degradation. Similar DMC systems are being tested in other parts of the world, including sub-Saharan Africa.

The new green revolutions

In this chapter, we have barely scratched the surface of the entire body of evidence on sustainable intensification now spreading across the small farms of developing countries. While much remains to be done, it is clear that remarkable progress is being made towards sustainable intensification of agriculture in many ecological and social contexts. Some of these are now occurring on large areas, with farmers and community groups pushing forward the innovation frontier. In these cases, sustainable intensification is no longer a minority exercise. What we need now are supportive policies that help farmers take advantage of the knowledge already created.

As we have shown, some cases of adoption at scale have taken place precisely where social and human capital development have intersected with helpful policies. Sustainable rice (and crop) intensification is one example. Agroforestry in sub-Saharan Africa is another. Yet, in many cases, progress has occurred despite, rather than because of, supportive policies. Farmers and community groups still find themselves defending niche innovations, in the face of incumbent regimes using conventional inputs and resource-intensive methods. The myth that smallholder, peasant systems have 'failed', or need constant and expensive interventions, is remarkably resilient. But as we have shown here, smallholders across the developing world are truly leading a new revolution for people and planet.

6

SUSTAINABLE INTENSIFICATION IN INDUSTRIALISED COUNTRIES

Over the second half of the 20th century, agriculture in the industrialised and affluent countries has become more productive, producing landscapes with larger and fewer farms, yet then also resulting in harmful side-effects that have economic consequences. Some of these side-effects have been directly on the environment, some on public health as a result of changes in food consumption or contamination by pesticides and pathogens. Not all these changes arise directly from agriculture, nor will making agriculture more sustainable entirely solve them all. As in developing countries, we see that the problems of the food system go beyond the growing of crops or the raising of animals – though challenges that begin here radiate across food chains. Solving the problems of unsustainable production will not, by themselves, guarantee enough nutritious food, at fair prices, for all. But it is a start, and will go some way towards meeting the complex challenges of food and nutritional security amongst the affluent consumers of the industrialised world.

In this chapter, we focus on the innovations and redesign occurring in some agricultural systems. We will show that there have been great advances on efficiency, some on substitution at scale, but relatively few on landscape and agroecological redesign.

Progress on efficiency and substitution

Farmers across industrialised countries have been experimenting with new ways of stewarding farmland over much of the 20th century. Some initiatives have sought to reawaken traditional practices, reimagining them for the modern age. Others have developed and deployed entirely new innovations to improve both sustainability and productivity. Some alternatives have developed in response to warnings on the environmental and social costs of conventional agriculture, as well as in response to the call for a more positive relationship between people and the planet. Some focus solely on improved efficiency, or on mitigating particular negative externalities. Some seek a redesign of agroecosystems, and the social fabric with which they interact.

With productivity improvements over the past 50 years, as shown in Chapter 2, farmed landscapes in industrialised countries have seen remarkable transformations. These have mostly been achieved on existing farmland, rather than expansion into natural habitats. At the same time, industrialised landscapes have lost good agricultural land to urban expansion and road infrastructure: in the EU-28, total agricultural land has fallen by 14 per cent to 188 M ha, including a fall in arable from 128 to 108 M ha (Buckwell *et al.*, 2014). The response to emerging evidence on the harm caused by agriculture to environmental services and human health was primarily focused on efficiency gains: how to reduce losses of nutrient, water, pesticide and farm waste. Crop breeding produced varieties more efficient at converting nutrients to biomass, also varieties resistant to pests and diseases. Advances were also made on the more targeted application of fertiliser and pesticide compounds, producing an approach now called precision farming (Garbach *et al.*, 2017).

Today, precision farming incorporates sensors, GPS mapping, information technology and accurate delivery systems to ensure inputs are applied in the correct place at the right time and dose. Drone mapping from above is now helping in the early identification of crops needing irrigation, nutrient or pest control. Energy efficiency has been increased through automatic control of agricultural vehicles using satellite navigation: this reduces overlaps of application. Controlled Traffic Farming (CTF) also uses satellite tracking to ensure machinery keeps to specific routes to limit the area affected by soil compaction. Through these methods, farmers benefit by spending less per unit of output, and the environment benefits by being less polluted. It is in these efficiency areas that robotic agriculture will make its first mark in the years to come.

None of these innovations, though, guarantee that the most optimal outcomes for land, farming and the environment will be achieved. They represent welcome steps toward sustainable intensification, and are necessary but not sufficient conditions for success. They may also require large investments by farmers in new technology: CTF, for example, requires a suite of machinery with matching axle widths, but there is a limited range of commercially available equipment that meets this emerging need. At the same time, biotechnology through genetic modification or gene editing may be able to produce new varieties with better resistance to disease and pests. If farmers can access and afford such new technologies, then we can expect that some substitution will follow.

As we have shown in the analysis of integrated pest management projects in Chapter 5, about a half of all pesticides used in farming are not agronomically necessary. In some agroecosystems, this can rise to 100 per cent. A recent review of nearly a thousand commercial farms in France found that on 75 per cent of the farms, low pesticide applications did not lower productivity or profitability. It was estimated that total pesticide use on these farms in industrialised landscapes could be cut by 42 per cent without compromising yield (Lechenet *et al.*, 2017). But despite these opportunities and advances, and thus greater yields per input (whether nutrient, pesticide or fossil fuel), existing agricultural systems remain significant contributors to stresses on planetary boundaries (Campbell *et al.*, 2017).

Nitrogen and phosphorus are ecologically important contaminators of surface and ground water, much derived from overuse of inorganic fertiliser and the losses of nutrients from farms in animal waste (Conway and Pretty, 1991). Precision farming can reduce losses from fertiliser, and appropriate waste management facilities can prevent pollution incidents. In the UK, a third of diffuse water pollution is caused by agriculture, comprising nutrients, faecal bacteria, pesticide compounds and soil sediment. Local water quality can be greatly improved through better soil management, planting of conservation field margins and the use of cultivation of nutrient soaks, such as late mustard. All are designed to intercept pollutants and contaminants. In addition, through redesign of crop rotations, however, farmers may be able to reduce or eliminate the use of some fertiliser. Wheat grown after legume yields more than wheat grown after cereal (Angus *et al.* 2015), and similar effects are noted for oilseed rape and potatoes grown in rotation with legumes (Charles and Vuilloud, 2001). Legumes can also have positive effects on soil health and crop resistance to pests and disease, and can also support biodiversity on and around farms, providing cover for small mammals, birds and insects and forage for wild and domesticated pollinators.

A recent study of lowland arable farms in the UK using 13 years of data illustrates the acute challenges of optimising very different yet desirable farm and environmental outcomes (Field *et al.*, 2016). Creating wildlife habitats from 10 per cent of the land combined with increasing rotational diversity increased breeding birds by 177 per cent and reduced greenhouse gas emissions by 10 per cent, but also reduced food energy production by 10 per cent. A more modest loss of yield of 1 per cent could still increase bird numbers by 50 per cent, though greenhouse gas emissions would only fall by 1 per cent. Such findings add further to the sense that there is no single transitional pathway towards sustainable intensification, nor single outcome or objective more important than all others (Chantre and Cardona, 2014; Hill, 2015). A review of the sustainable intensification of European agriculture concluded that the environmental outputs of land management should be on an equal footing with food and energy outputs (Buckwell *et al.*, 2014).

In European agricultural systems, supported by payments under the most recent revisions of the Common Agricultural Policy for the period 2014–2020, these kinds of environmental mitigations have become more common. On average, Pillar I and II payments comprise 50–60 per cent of farm income, justified on the grounds that public money is being used to protect or improve public goods. Pillar I payments are direct payments (€25 billion over six years) that require farmers to adhere to EU regulations, with 30 per cent of the payment dependent on each farmer implementing greening measures: these cover crop diversification, 5 per cent of farmland dedicated to environmental benefits (ecological focus areas) and measures to maintain permanent grassland. Organic farmers automatically qualify for the greening payment. Pillar II payments (€2 billion) provide benefits through voluntary agri-environment scheme (AES) measures with responsibility devolved to national governments: in England, the focus is on woodland creation, biodiversity and water management. The effectiveness of the greening for enhancing biodiversity under Pillar I has often been questioned; AES under Pillar II are widely thought to be very effective. In the UK, it has been estimated that farmers created £1 billion of net environmental benefits as a result of receiving public subsidies and support for ecological focus areas (EFAs) (POST, 2017).

One drawback to all these support measures is that they work almost solely at the individual farm level, rather than seeking positive and synergistic changes across whole landscapes. As we shall see in Chapter 7, the element of social capital creation is an important prerequisite to transformations to sustainable intensification at scale. In the USA, some of the best examples of improvements to environmental services have come through locally developed initiatives around farmer-managed watersheds (see below); in Australia, the National Landcare Programme has had widespread and positive impacts on farm practice and environmental services.

Agroecological methods for redesign

As we have indicated above, farm redesign has been less commonly seen or experienced in industrialised countries, save for organics and small-scale community-supported agriculture. There have been initiatives to encourage whole farm and environment thinking and planning, including the many efforts to develop indicators to measure sustainability, but these again tend to result in changes at the efficiency and substitution end of the ESR spectrum.

In 1997, one of us presented this agroecological vision in the first use of sustainable intensification (Pretty, 1997): it should be an agriculture:

> relying on the integrated use of a wide range of technologies to manage pests, nutrients, soil and water. Local knowledge and adaptive methods are stressed rather than comprehensive packages of externally supplied technologies. Regenerative, low-input agriculture, founded on full farmer participation in all stages of development and extension, can be highly productive.

There are many advances to report in industrialised agricultural systems around the use of agroecological principles. Intercropping substitution in the western Canadian prairies has recently been shown to be effective: lentil and dry pea legumes replacing spring wheat in a four-course rotation reduced environmental impacts, reduced nitrogen use and increased returns to farmers (MacWilliam et al., 2014). This represents a small step towards redesign. Multiple interventions have been shown to improve water quality in the US Midwest, with composted animal manures, use of forage legumes and increase rotational diversity together raising water quality on and near farms (Cambardella et al., 2015). In the UK, beetle banks of tussocky grasses encourage ground beetles that move into crops to control aphids, and conservation headlands typically 2–6 metres wide save on input costs but also encourage populations of beneficial insects (Bianchi et al., 2006; Lampkin et al., 2015).

TABLE 6.1 Some agroecological applications in UK agricultural systems

- Crop and livestock management
- Catch crops and green manures
- Winter cover crops, bird food crops, over-winter stubbles, arable in-field bare bird-patches
- Conservation headlands, beetle banks and flowering field margins and strips
- Crop residue management
- Diversified crop rotations, including polycultures (e.g. spring oat and beans for home-grown high protein feed), and diverse grass swards (e.g. grass and trefoil, chicory and plantain)
- Intercropping, including legumes in animal (e.g. sheep) systems
- Alley cropping and leguminous trees
- Zero or reduced tillage
- Integrated pest management, reduced pesticide use
- Use of traditional livestock breeds
- Optimising manure and waste applications
- Integrated animal and crop systems, with a focus on grassland ecology
- Mob grazing and rotational grazing (management-intensive grassland management).

Some innovations developed in the tropics have been applied: push–pull principles have now been used for the first time in the development of trap crops of turnip rape that attract pollen beetles out of oilseed rape. Others, though, have been harder to transfer: no tillage in many temperate systems is harder to make effective because of weed and slug damage. Table 6.1 lists the range of agroecological innovations that have been used in UK systems. In most countries, few agroecological innovations have gone to scale: the exceptions are in Germany and France, where governments support over 600 projects (Lampkin *et al.*, 2015).

Organics as redesign

The centrality of ecological processes in agriculture was recognised by F.H. King in his 1911 book, *Farmers of Forty Centuries*, in which he wrote of traditional farming agriculture in China and Japan maintaining highly productive systems by cultivating biologically rich soils. Later, Albert Howard, observing peasant farmers of India, wrote of the recycling of organic wastes back into the farm to supply nutrients and maintain healthy soil structure. The writings of King and Howard gained prominence alongside the work of Robert Rodale and Eve Balfour, though opposition to organics and regenerative agriculture was sometimes fierce: the movement was called cultish, just romantic nostalgia or even dangerous bunkum (Truog, 1946). Concern about lower yields in organic systems prompted one US Secretary of Agriculture, Earl Butz, famously to warn in the 1970s that a switch to organics would mean 50 million Americans would starve.

It is at this point that we see a divergence between what is sometimes now called conventional agriculture and the range of alternatives. Creating dichotomies can sometimes seem helpful (identifying the goods and bads of systems), but has tended to prompt oppositions and dismissals based on emotion rather than fact. This has occurred on both sides: conventional farmers dismissing organics, and organic or alternative farmers dismissing equally those who use artificial or manufactured pesticides and fertilisers. This has not always helped in the development of solutions to common problems felt by all. John Reaganold and Jonathan Wachter (2016) have recently called for advances beyond these dichotomous narratives of either–or choices of language and technologies: they have called for a *beyond organic* or *organic 3.0*, with a focus on blended systems. Lampkin and colleagues (2015) have also noted there is a range of constructive groups that now recognise the values of both agroecology and technological approaches.

By the 1970s, the first standards for defining organic agriculture were developed in Europe and North America. Today, there are some 40 Mha of certified organic agriculture (Reaganold and Wachter, 2016). Yields are generally some 8–25 per cent lower than conventional equivalents, though under certain conditions organic yields can match conventional (Seufert *et al.*, 2012). Generally organic systems have higher biodiversity, landscape diversity and soil carbon, and lower soil erosion and contamination of water systems. As we showed in the previous chapter, yields on organic and resource-conserving initiatives in Africa and Latin America can increase: 80 per cent of 25 cases showed both food and net incomes growth in organic systems (Bennett and Franzel, 2013).

In Europe, some 2.5 per cent of agricultural land is now under organic production (Willer and Lernoud, 2016). Over the past decade, the number of organic producers has grown by 55 per cent and organic area more than doubled. At present, just over 11 million hectares of land across the European Union is cultivated organically (IFOAM-EU, 2017; Council of the European Union, 2017). In the USA, 1 per cent of maize, wheat and soybean are grown organically, with just under 15,000 farms certified organic in 2016, covering some 1.6 Mha (TakePart, 2016; USDA, 2017). Most is permanent pasture, then cereal and other permanent crops.

Part of the attraction of foods and other produce derived from organic systems lies in the promise of co-creation of environmental and health benefits (Tuck *et al.*, 2014). Organic products have generally had price premiums added. This may put them out of the range of affordability for some consumers. It has also been shown that organic systems internalise some environmental costs by not causing pollution (which then costs to clean up): farmers receive a benefit for this from consumers through prices, but not from governments for creating or protecting public goods (Pretty *et al.*, 2005).

The health implications of organic foods remain discussed and often contested: some say organic foods taste better as well as are beneficial to health. Comparative studies have focused on both nutrition content and lack of contaminants, pathogens or pesticide residues; some have been designed as field trials, basket studies or farm surveys (Barański *et al.*, 2014). The health impacts of particular components of the diet are also complex, with individuals responding differently based on levels of exposure, food preparation techniques and general health. A number of meta-analyses have sought to pull together large numbers of studies and attempted synthesis. Dangour *et al.* (2009) compared the nutritional content of organic and non-organic foods, and found few significant differences. A later study compared health effects and concluded that organic foods may not be more nutritious, but may reduce the risk of exposure to pesticide residues and antibiotic-resistant pathogens (Smith-Spangler *et al.*, 2012). More recently, a widely publicised meta-analysis of 343 peer-reviewed studies (Barański *et al.*, 2014) compared nutritional content and found that organic foods contained significantly higher concentrations of phenolic acids, flavanones, stilbenes, flavones, flavonols and anthocyanins, all of which have been linked to beneficial health outcomes including reduced risk of chronic disease, including cardiovascular disease, neurodegenerative diseases and certain cancers. Non-organic foods were also found to contain pesticide residues four times more frequently, as well as higher concentrations of cadmium.

An important and somewhat open question, then, centres on how consumer choice may influence the spread of sustainable intensification. Consumers eat every day, and vote every few years. Their choices shape agricultural systems. Here labelling matters: organic and conservation-grade systems in the UK have been shown to have greater diversity of habitats and species than conventional systems (Tuck *et al.*, 2014; Hardman *et al.*, 2016). In France, a rising demand for organic pasta has encouraged the establishment of new participatory durum wheat breeding initiatives (Chiffoleau and Desclaux, 2006). Farmers, millers, bakers, advisers and consumers together explored different landraces of wheat, conducted on-farm experiments, and helped farmers choose the best wheat varieties for different products. In the USA, high consumer demand for organic food in the cities of California and New York means that these states top the country in organic acreage. But much remains to be done, and there is tremendous potential. Consumers in North America and Europe together generate 90 per cent of the world's organic food and drink sales, but much of the organic food they consume is grown elsewhere (Sahota, 2016).

There may also be further influences on the social capital between farmers. Farmer seed networks across Europe and North America have engaged networks of farmers and other stakeholders to preserve varietal diversity for a number of commercial and non-commercial crops. Networks such as Réseau Semences Paysannes in France, Red de Semillas in Spain, and the Rete Semi Rurali in Italy bring together farmers and scientists, working together on participatory breeding and varietal improvement. Participatory breeding has been particularly central to organic cultivation, where farmers are often challenged by heterogenous growing conditions and a lack of suitable crop varieties (Chiffoleau and Desclaux, 2006).

The case of the Cholderton Estate, UK

The Cholderton Estate comprises 1,000 hectares on the Hampshire–Wiltshire border in the UK. It exemplifies modern highly competitive agriculture with the preservation of countryside features and natural capital. Under the leadership of farmer Henry Edmunds, it seeks to "farm in harmony with the environment" through mixed farm enterprises, organic methods, dairy, rare breeds and restoration of chalk grassland. It has implemented a carbon reduction programme, and now verges on neutrality. Grassland is central to the livestock operation: the dairy herds, Hampshire Down sheep and Cleveland Bay horses graze legume leys most of the year, and the estate is self-sufficient in all livestock feed. Barley, oat and vetch are grown for animal feed: no weed control is needed. The grass-legume (red and white clover, lucerne, sainfoin) leys are productive, important for pollinating insects and ground-nesting birds, especially lapwing, skylark and corn bunting. The Cholderton and District Water Company provides clean and reliable potable water to the estate and 2,000 people living in villages over 21 square kilometres. The organic systems of production provide this service of clean and uncontaminated water.

The contrary farmers in Ohio, USA

Ohio farmer and writer, Gene Logsdon, has written about being a contrary farmer: small, biodiverse, community-oriented, unstressed and above all successful (Logsdon, 1994, cited in Pretty, 2014): "We are pioneers, seeking a new kind of religious and economic freedom." Such contrary farming depends on reducing labour to a minimum by skill instead of using expensive machines. His 20-acre farm has 130 species of birds, 40 of wild animals, 90 of wildflowers and trees. The more diversity on a farm, the better the self-regulation. He does not spray: "it's not a problem". He writes that Amish farmers are geniuses: they produce at low, horse-power costs, and sell at high, tractor-powered prices. There is also something else about this farming style: "the era of horsepower was just as much fun and far less stressful than the high-tech days of later years".

He asked of one of his neighbours, why keep sheep? "Well, there's little money in it," she conceded, "but the real reason is my sheep make me happy." He is an adopter of rotational grazing, the practice that has transformed livestock production in recent years. "To understand a meadow," he writes, "you really need to sit down in one a while. Maybe like for twenty years." Farmers who pay attention, in short, are likely to be successful. "My animals notice more about me than I do," he notes, observing something important about the body language of farmers and their animals. "Thousands of us contrary farmers are the village idiots of agriculture. We farm because we like to make little paradises out of our land while growing good food on it." This means these farmers are able to ride out financial crises in the way that large indebted farms cannot.

Amish farmer David Kline is editor of *Farming Magazine* that covers the perspectives of small farmers. He practises rotational grazing, and farms only with horses (Pretty, 2014). The system of intensive grazing deliberately constrains cattle to small patches of pasture for short periods. They eat quickly, fill up, and make milk while resting. The fencing is moved each morning and noon. The cattle are thus encouraged to eat all plants, and not just their favourites. All the feed for 45 Jerseys comes from the farm; there are 50 pairs of bobolink nesting in the legume-rich pasture. The Organic Valley cooperative was originally set up by seven farmers, and has grown rapidly in the past decade: now the 1,600 members have a turnover of $650 million. Son-in-law, David, says, "in conventional markets, you're not used to the company batting for the small farmers, but here they do". The cooperative arranges third-party certification, manages quotas and prices, keeps the suppliers content. Rotational grazing now means farmers are having to select cattle with a strong, wide front-end, a large rumen, large mouths for grazing all types of pasture plants, and shorter legs. They yield less milk, but require less time for milking.

On the Schlabach farm near Berlin, there is also rotational grazing and intensive polytunnels. "Back in the 90s," explained Rob, "it wasn't going well for us. We were trying to push production to the maximum in a mix of conventional and Amish ways. We did not have mechanisation of conventional farms, and yet were trying to follow the same path." They made a redesign shift, taking up a smaller-scale system that lets animals do the work. Their 120-acre farm is mostly under grass. Polytunnels grow organic vegetables watered by drip irrigation. Onions and squash are on the top of one hill. A green lane is cooled by the shadows of trees. "This is all new," observed Rob. "I used to cut all the trees down to make use of every inch of the farm. Now we let nature do the work for us." He can remember the soil washing fast down lanes when rain fell.

Over the fence, a neighbour declared: "I'd have to have brain surgery to farm like you do." And now, he cannot find a way to swallow those words. On the organic side, the soil is deep, rich with organic matter and carbon. They too are redesigning the genetics of their animals, more towards animals more typical of the 1950s. The health of animals has improved too. Organic cattle produce about 6,000 kg of milk per year, only half that of conventional animals. There are savings on trimmings of hooves, as the cattle are out on the land. Barn owls have returned, taking to flying behind the cattle as they walk up mice and voles while grazing. The *produce-more, get-more paradigm* has changed.

These contrary Amish farmers note that it is the constant observation and management of pasture that makes them better grass farmers. Young people have come back into farming because of the high knowledge input required for the more knowledge-intensive rotational grazing. The next generation of innovation may be around anaerobic digesters and solar photovoltaics, but the knowledge base will also have to expand on integrated pest management, cover crops and soil amendments.

Small farms, patches and hydroponics

In the previous chapter, we showed how patch intensification in developing countries is an important part of many food systems, even though by definition these systems are small in area in landscapes. The same is increasingly true in industrialised countries. Progress towards sustainable intensification has occurred on small patches, often led by consumers and growers from non-traditional farm backgrounds. Home garden and allotment cultivation has long been an important part of the food system, yet also neglected. Now interest and demand is increasing, and this closer attachment to the land may further shape the adoption of sustainable intensification. In the UK, there are 20 million gardens, 330,000 allotment plots (Crouch and Ward, 1997; The National Allotment Society, 2014); the American Community Gardening Association (2014) estimates that there are 18,000 community gardens and 35 million people grow their own food. In many industrialised countries, an alternative food movement has now emerged, largely around small farm or garden systems that are often urban. It has been driven by discontent over the negative externalities of industrialised agriculture, lack of attention to animal welfare, concerns over food safety and desires for social justice around food. These have led to the emergence of a broad range of alternative forms of food production and provisioning which together constitute an attempt to shape food systems more directly (Little *et al.* 2010). Informal learning, experimentation and experience are at the heart of much urban agriculture, often run by non-professional, volunteer groups of cultivators. Here too there is a knowledge gap, where additional research is required. Many community gardeners are motivated by 'activist ideologies', are not professional farmers and run enterprises without specific training, and here, too, they are at the cutting edge of innovation.

In these ways, small patches are linked to consumers through community-supported agriculture (CSAs), box schemes, tekei groups (Japan), food guilds (Switzerland) and farmers' markets. In the USA, CSAs have grown in number from just 2 in the 1980s to 7,400 in 2015, with 50,000 consumers as members; the number of farmers' markets is up from 1,700 in the mid-1990s to 8,700 (Pretty *et al.*, 2015; USDA, 2016); and the number of farm-to-school programmes has risen from 6 in 2001 to 10,000 (National Farm to School Network, 2013). A total of 144,000 farms are involved in these direct sales schemes, generating more than $3 billion of income for farmers.

The greatest number of direct group consumer-to-farmer links are in France, where 1,600 AMAP groups (Associations pour le maintien d'une agriculture paysanne) comprising 50,000 families are supplied with weekly vegetable boxes, benefiting 270,000 individuals. The focus is wider than group purchasing and quality of food: AMAP farmers engage in agriculture biologique, and there is explicit attention to the value of retaining small farmers in the landscape (AMAP, 2017).

In the UK, half of consumers surveyed by the New Economics Foundation expressed an interest in buying locally grown food (NEF, 2003) and around 90,000 are awaiting allocation of allotments on which to grow their own. It is now estimated that five forms of community food initiatives, comprising more than a thousand individual enterprises, such as community-owned shops, farmers' markets, country markets, CSAs and food co-ops, produce a combined annual turnover of £77 million. SERIO (2012) estimated that the total economic value of community food enterprises in the UK is £150 million per year. One of the oldest US CSAs is Angelic Organics in Illinois: it provides food boxes to Chicago. They say, "farming is not just about raising vegetables, but also about those who receive the vegetables. To reunite consumers with their source of food, to share the magnificent drama of farming". This suggests modes of personal, social and ecological redesign (Hill, 2014).

Private gardens may also make a significant contribution to household food supply. In the late 1950s, it was estimated that 14 per cent of private garden area in London may have been allocated to fruit and vegetable cultivation (Wibberley, 1959). Though the current figure may be lower, in 2000 it was estimated that assuming productivity of 10.7 t/ha and consumption of 0.5 kg of fruit and vegetables per capita per day, London could produce enough to supply its residents with 18 per cent of intake (Garnett, 2000). In Chicago, the aggregate production of home gardens may even exceed that of community gardens and other forms of urban agriculture (Taylor and Lovell, 2012). In Florida, it was estimated that small gardens yielded 69 per cent of vegetables consumed by farm families (Gladwin and Butler, 1982).

Transformation and redesign of city space has also started to occur. It has begun to happen in forlorn Detroit; once the capital of cars, the population fell from 1.9 million in the 1950s to 700,000 today (Detroit Food and Fitness Collaborative, 2014). It has more than 30,000 deserted plots and vacant houses, taking up 40 square miles, nearly the size of San Francisco. But the city government has few financial resources and, with industry in freefall, has few prospects. The city does not have a single national grocery chain in the city: it is as close to a food desert as can be imagined. Yet urban farming has taken root, and 1,400 small farms and community gardens have been formed. Some are run by families with no work, some by community groups. Some such as Hantz Farms are large operations intending to create commercial city farms that would bring jobs, something that has eluded economic planners year after year.

But setbacks can follow advances. In the 1990s, 700 GreenThumb gardens emerged to great success in New York City, but then were swallowed up by developers. This may happen in the former rust-belt cities, but at least not for some time. There are many institutional models: community gardens of orchard trees and vegetables that make all the produce available free to locals. Such gardens are rarely vandalised. Rich Wieske runs Green Toe Gardens, an apiary with 60 beehives across Detroit. "There is so much forage now," he says, "so much land for bees." They sell 1,500 kg of honey annually. Some 900 food gardens have now appeared in Detroit. One retired truck driver shoots racoons and sells them for $12 apiece. Pheasants have become abundant, also favoured for the pot. A wild and agrarian city is emerging. It will have low food miles, a small environmental footprint. But not everyone is in favour. Some have declared "this is a city, not a farm", and are waiting for the revival of the old industrial model. Others are disturbed by the agrarian vision: there is green where there should not be.

A further example of green in the city is provided by the emergence of small-scale hydroponic production systems. These grow crops without soil, using direct delivery of nutrients in water. In Japan, rice is being grown hydroponically in underground vaults, providing four full cycles annually. In Israel, where agricultural land and water are in short supply, a range of hydroponic systems are under development. In the USA, there are 2,300 hydroponic farms providing mainly tomato, lettuce, herbs and cucumber. Revenues in 2016 were just over $800 million, growing at 4.5 per cent per year (Ali, 2017). One single hydroponic farm in Connecticut will cover 4 hectares, generate 40 jobs and produce 800 tonnes of tomato per year (equivalent to 200 tonnes per hectare).

The Goat Lady Dairy, North Carolina, USA

In Randolf County, North Carolina, Steve Tate and family run the 24-hectare Goat Lady Dairy. This is a landscape where cotton and tobacco once were dominant, and where farmers today struggle to make a living from these commodities. Steve, though, had a different kind of concept for his farm, some three-quarters of which is woodland: "we started by saying, what do we want to do every day when we wake up?" Their dairy produces a variety of distinctive hard and soft cheeses from the herd of 30 goats. They run regular dinners-at-the-dairy, where all the served food is from the neighbourhood, and altogether 2,000 people visit the farm each year. Steve says "we tell the story about the whole farm – never underestimate the desire of people to get in touch with the land". This is a small farm in a struggling farmed landscape: the family says happiness is the most important thing to them.

Management-intensive rotational grazing (MIRG) systems

Extensive low-intensity grazing systems are well adapted to many landscapes, such as the dry steppe of Asia, savannah of Africa, tundra and boreal habitat of the sub-Arctic, rangeland of North America and wet upland of Britain. Cattle, sheep, goats and reindeer are important shapers of whole landscapes as well as sources of food. In these contexts, intensification is not desirable, mainly because the landscapes are generally at or close to carrying capacity. However, there has been a recent rapid expansion in intensive grazing management systems, particularly in the lowlands and on smaller farms (NRC, 2010). These management-intensive rotational grazing (MIRG) systems (also known as mob grazing) use short-duration grazing episodes on small paddocks or temporarily fenced areas, with longer rest periods that allow grassland plants to regrow before grazing returns.

These systems substitute knowledge and active management for external inputs to maintain grassland productivity. As many have replaced zero-grazed confined livestock systems, the animals themselves are bred for different characteristics: large mouths, shorter legs, stronger feet and hooves, larger rumens. MIRGs were first developed in New Zealand, and now have become common in parts of the USA: on 20 per cent of dairy farms in Wisconsin, Pennsylvania, New York and Vermont (NRC, 2010). Some whole communities, such as Amish of Holmes County, Ohio, have converted all dairy systems to rotational grazing (see the contrary farmers case above; and Pretty, 2014), where their response to family labour availability and reduced costs of MIRGs has been to reduce animal milk productivity to reduce the time needed at milking.

There is good evidence that MIRG systems result in improved soil quality, increased carbon sequestration, reduced soil erosion, improved wildlife habitat and decreased input use (Undersander *et al.*, 2002; Hensler *et al.*, 2007). Livestock in

TABLE 6.2 Economic indicators of performance of three systems of dairy production in Wisconsin (mean over 8 years)

	Management-intensive rotational grazing	*Traditional confinement system*	*Large recently developed confinement system*
Kg milk per cow per year	34,570	43,070	49,510
Costs ($) per 1,000 kg of milk	165	170	180
Net farm income ($) per 1,000 kg milk	69	47	51

Source: NRC (2010).

housing create waste disposal challenges and costs; livestock continually graze and manure the land. But animals on the same grassland for too long cause overgrazing, sparse pastures with low persistence and soils low in carbon. In typical MIRG systems, animals are moved twice a day. This requires high levels of monitoring and active management by farmers. Such short and intensive periods of grazing mean the animals consume all plants, rather than leave those they otherwise find unpalatable. Well-managed grazing systems have been associated with greater temporal and spatial diversity of species. There is also evidence that MIRG systems result in greater animal welfare (Fuhlendorf and Smeins, 1999). MIRG systems outperform traditional confinement (50–75 cows) and large modern confinement systems (250 cows) on economic measures (Table 6.2). It is estimated that there are ten thousand farms in the USA practising rotational grazing on some 1.6 Mha.

Landscape redesign

A key challenge for sustainable intensification is the delivery of transformation at landscape level. Farmers can change individual practices on fields, but ecological and social shifts occur when whole landscapes are redesigned either by collective action or by the influence of policies. A good example is the influence of the European regulation to preserve biodiversity across farmed and non-farmed landscapes. Together, the Birds and Habitats Directives set the legal basis for a protected areas network, Natura 2000, the world's largest consolidated network of nature conservation areas. Much of the land protected by Natura 2000 is farmland and mixed semi-natural habitats such as hay meadows and grazing lands. Additionally, under the Common Agricultural Policy, Pillars I and II of agri-environmental schemes now directly support farmers who commit to stewarding their farms for environmental as well as farm outcomes, though does not guarantee synergies between farms.

In the UK, the institutional hole left by the closure of the public extension system in the 1980s has not been filled. Some charitable organisations, such as LEAF, have been able to develop non-spatial networks to share good practice, but these do not create synergies in landscapes. As indicated with the case above of the Cholderton Estate, some individual farmers have large enough enterprises to undertake landscape-scale actions as sole operators. The first national effort to create spatial initiatives was the Demonstration Test Catchments established by Defra. This brings together various actors around Sustainable Intensification Platforms (SIPs) established as outdoor laboratories. The focus is on reducing the impacts of nutrient, sediment, microbial and pesticide contaminants and pollutants. Four platforms have been established in the catchments of the rivers Eden, Wensum, Avon and Tamar, and farmers are involved in the co-design of experimental mitigation measures.

Another small but important innovation also centres on the need to engage farmers in research (MacMillan and Benton, 2014). The Duchy Originals Future Farming Programme, led by the Soil Association and Organic Research Centre, has organised supplier farmers into groups of 5–15 farmers, who then jointly tackle a problem put forward by a participant. Solutions are tested for up to a year, and jointly evaluated. Some 450 farmers have now been involved in these field labs. These bring farmers together, and while they do not necessarily provide clear answers, they do set the scene for collective action towards redesign. It is interesting to note that this institutional innovation comes far from the centre of UK agriculture: an organic NGO, a non-publically supported research centre, with funding from a charitable foundation.

The case of Elmley wetlands: wildlife-friendly farming in the UK

The draining of wetlands was long a key strategy during early phases of agricultural intensification, yet their demise also meant the loss of important environmental services. When coastal wetlands were removed, wader bird numbers declined and flood incidents increased. At Elmley Marshes on the north Kent coast, 1,200 hectares have been farmed for the past 40 years by Philip Merricks for both rare waders and farm produce. Each year, some 15–20,000 visitors come to the Elmley farm. It was the first working farm in the UK to be declared a National Nature Reserve, with visitors attracted by both landscape and birds. Each winter, 30,000 wildfowl and waders stop over, filling the water and sky with their swirling presence. By spring, the lapwings and redshanks are breeding, along with oyster catchers and iconic avocets.

The management design is thoughtful and precise, as birds need different microhabitats, including wet puddles and rills, scraped earth, low grass, taller grass in which to hide, and both dry and wet patches. Winter waterlogging attracts wigeon and dunlin, also slowing spring grass growth, thus allowing young lapwing to survive. Predators have to be controlled: if breeding waders are favoured, then fox and crow have to go; though heron, falcon and gull will also take chicks and eggs. Breeding wader numbers have increased tenfold on the farm over the past 20 years. Philip Merricks says this remote and open landscape has "an elusive and powerful charm. Our birds are our landscape". Much can be done when the wild and agriculture are designed together.

Peanut farmer groups, North Carolina, USA

North Carolina peanut growers had come up hard against an economic barrier. Peanuts are important in North Carolina: 2,300 farmers produce 170,000 tonnes per year, the fourth largest amount by any state in the USA. Since the 1930s, the Federal Peanut Programme had maintained a steady and predictable price, with prices elevating whenever costs increased. But in the mid-1990s, the programme was radically changed. Prices were cut and quota carry-over eliminated, resulting in dramatic falls in farmer income.

Out of the crisis came collaboration. With the help of Rural Advancement Foundation International, a group of 62 farmers began reinventing both local farming and social relations. Over a period of four years, these farmers reduced pesticide use by a remarkable 87 per cent, saving themselves $40–50 per hectare in costs without any yield penalty. On more than three thousand hectares, they had managed to cut pesticide use by 48,000 kg. The change in attitudes and values was rapid: this was redesign in action. A major pest of peanuts is thrips, yet most leaf damage has no yield effect, even though the crop looks damaged (similar to the first 40 days of leaf damage in irrigated rice systems in Asia). By conducting their own research, farmers came to realise they did not need to spray: "we were farming for looks", said Rusty Harrell. Michael Taylor added, "the peanuts did not look good – but the yields increased".

The key to success was scientific experimentation by farmers and peer-based learning. Farmers set the agenda for field trials of alternative practices, and watch for unexpected results and are encouraged to be careful about drawing conclusions. Working together, sharing experiences and developing new relationships of trust are central components of the process. "We got together over food, and found we had a wide range of problems, and were all searching for new ways", says Rusty. "We go around and look at other people's crops." Farmers in the group indicate this helped to bring the community together. Importantly, there are no end-solutions, as sustainable intensification needs continuous experimentation. Tom Clements said, "this has affected our lifestyles. I'm still working on it – you have to farm true every day. Our quality of life has improved." The field trials gave farmers the confidence to try something new, and the trust and sharing helps them to take large steps into the unknown. As a result, incomes go up, and the environment benefits too.

Farmer-led watersheds, USA

The USA has a history of addressing soil erosion at watershed system level since the establishment of the Soil Conservation Service in the 1930s. Recently this has involved the close engagement of farmers so as to achieve beneficial outcomes for both productivity and natural capital. The most famed example is the investments upstream of New York City's potable water supply to ensure dairy farmers in the Catskill range were able to produce water from their farms free of sediment, nutrient, pesticide and Cryptosporidium (NRC, 2010). The investment principle centred on upstream expenditure to create systems that produced clean water, rather than building plant to clean up water after it had been polluted. The former is a social project; the latter an engineering one. Interventions have tended to centre on efficiency and substitution rather than redesign, and thus may not guarantee farmer compliance over the long term.

The evolution of farmer-led watersheds has, however, created platforms for engagement and co-creation of locally based technology and practices to address specific challenges. In Wisconsin, Farmer-Led Watershed Councils involve university researchers, the Department for Natural Resources, Land Conservation departments of counties and the Wisconsin Farmers Union (FLWC, 2015). The principle problem addressed is nitrogen and phosphorus loss, which is costly to farmers and damaging to surface water systems that tended to be oligotrophic and nutrient-free. At the far end of many Midwest US watersheds is the Mississippi basin and Gulf of Mexico, where synthetically sourced nutrients have created dead zones at sea.

Farmers are engaged in analysis, design, monitoring and implementation of water harvesting methods. They are offered financial incentives and compensation to experiment with a suite of technologies: cover crop trials, nitrate testing in maize, nutrient planning, manure spreader calibration, phosphate indexing and intercropping. The initiative focuses on the St Croix and Red River Basins, on a range of watersheds each of about 8,000 hectares in size. Many farmers have changed practices completely. Brad Johnson has been farming for 40 years, and has adopted zero tillage for maize and soybean: it looks messy, but it saves time and money. The monitoring station at the farm boundary showed that there was no nutrient loss in run-off. Local lakes Apple and St Criox suffer severe algal blooms in summer, and now farmers are working together to reduce phosphorus discharges from their farms to restore water quality. The one thousand farmer-led watershed groups across the USA are thus a basis for redesign of agricultural systems and landscapes.

Landcare, Australia

The Landcare movement emerged in Australia in the 1980s following two to three decades of improvements to both productivity and environmental care: clover leys had introduced nitrogen, new varieties improved yields, erosion seemed to have been stopped. Yet mechanisation enabled large areas of bush, scrub and forest to be cleared at unprecedented rates. The negative consequences of intensive cereal and livestock farming became apparent, particularly as salinity of soil and water spread, and tree die-back became common. The National Landcare Programme (NLP) was launched in 1989 following a number of revegetation initiatives, many at the time imagining "a sylvan future with a thriving revegetation movement and vibrant farm sector" (Campbell, 1994; Campbell et al., 2017). Critically, both the Australian Conservation Foundation and the National Farmers' Federation joined forces, promoting a collective vision and consensus politics. To rehabilitate landscapes for multiples users requires people to work together, making such redesign an act of social ecology (Hill, 2014, 2105; Wright *et al.*, 2011). The One Billion Trees programme was adopted to encourage more tree planting, but also to provide hope, particularly for young people. The First National Treefest attracted more than 6,000 people to investigate tree planting and propagation.

With the key use of some national funding for rural facilitators to work as change agents, more than 6,000 Landcare groups were formed across Australia. Some focused solely on farmers and the innovations they would develop by working together; others focused just on conservation and species objectives, such as protecting turtles or frogs. More, though, sought and co-created both environmental and agricultural outcomes. Now Landcare looked like one of the most significant social movements centred on the land in any industrialised country: one-third of all Australian farming families had become actively involved in these 6,000 voluntary groups. Some transformations are personal too (Box 6.1).

BOX 6.1 The cotton women

Dalby is a small linear town on a distant crossroads on the Darling Downs in eastern Australia. Like many similar rural settlements, the community and economy is under threat. Farming seems to get tougher year on year, businesses struggle, children must travel further to school, and existing associations seem tired and inappropriate for the challenges that modern society brings. Cotton is big on the Downs, but farmers were struggling against growing pest resistance and increasing environmental degradation. Farmers organised to share ideas and practices for novel pest management, and 350 families joined the growers' association. There is progress towards sustainability. On the Jimbour Plain, Carl and Tina Graham reflected on the changes, by saying, "ten years ago, if you saw your neighbour spraying, we'd go out and do it too". Now they were scouting the fields, using trap crops, managing beneficial insects and using natural viral pesticides. A mosaic landscape was being created, so that sorghum can build up parasites, or lucerne benefit the cotton. They say, "we still have a lot of learning to do".

But something else was changing too, and that was inside people: it involved relations between men and women. The Women in Cotton group in Dalby was led by Catrina Walton, and had 60 members. They convened to talk about the pesticides used in cotton cultivation. One said, "we found it so powerful, just to get us all together". They meet once or twice a month, sometimes for discussion, or to hear talks from external professionals of their choice. They organise farm visits for several hundred children each year. The benefits of the group seem to centre on two things. The first is the value of the meetings. Said one member, "you feel safe, you don't have to tread carefully with your words"; and another, "social networking is one of the greatest things I get out of this group".

There were also changed relations within families. Women say they did not know enough in the past to ask sensible questions, and mothers tended to be pushed into the background. But now there is greater understanding in families, improved communications and more joint decision-making. One said "it makes for a better marriage". Carl Graham said, "when I come home from the paddock I get asked heaps of questions, and we interact more". Tina said, "women feel more involved. Now I have ideas for improvement, and can answer questions". The women themselves are adding productive value to the system. They read reports, help with marketing, and learn about pests and predators. Men tend to lack the social networks that women develop, and these networks help to spread good ideas. But it is not easy. One male agronomist arrived at a meeting to give a talk and, in front of 50 women, said, disappointed: "Oh, so there's no one here yet." Together, though, women and men are slowly redesigning their farm systems, making them more sustainable and productive, and they are doing so by crossing a multitude of personal, family and community frontiers.

Yet now, backward steps were taken: many groups became obsessed with competing for limited national and state funding, some suffered burn-out, later governments permitted greater bush and forest clearances, some switched to maintaining remnant vegetation, as stopping losses gained priority over creating new natural capital (Cary and Web, 2000; Curtis *et al.*, 2014). Engineering programmes were adopted to address problems after their cause, such as the $13 billion spent on water reforms in the Murray–Darling Basin. Private land conservation began to grow through the efforts of NGOs and charities, creating pockets and reserves rather than transforming whole landscapes. Despite these slippages, values have changed, hearts and minds have been won over towards conservation and social capital created that has had long-standing cultural benefits. Still some 4,000 community Landcare groups are active, with 60 per cent of farmers in broadacre and dairy sectors still actively involved. Andrew Campbell, James Alexandra and David Curtis (2017) recently concluded, "the countryside looks in better health today than it did forty years ago".

In 2017, the new coalition government agreed to invest $100 million into the National Landcare Programme, seeking to see 20 million new trees planted by 2020. The NLP is focused on maintaining and improving ecosystem services, increasing the number of farmers and fishers adopting practices that improve the quality of the natural resource base, increasing engagement and participation of whole communities, and increasing restoration and rehabilitation of land. It is conceivable that the Landcare movement, with this new investment and central support, may now enter a new and positive phase.

Advances in the affluent countries

As we have seen, there has been some encouraging progress towards sustainable intensification on the larger farms of the affluent and industrialised countries. Though long legacies of unsustainable cultivation persist, there are many shifts towards greater efficiency and substitution, and some examples of redesign. An encouraging upwelling of social capital has been at the heart of these shifts, as have some supportive policies. Farmer groups, landscape-scale collectives and crucially new communities of consumers who wish to re-engage with the land have all driven forward the edge of innovation.

We now turn our attention in the next chapter to the critical factor for all sustainable intensification: social capital creation, development and main-tenance. To us, the evidence in the last two chapters shows clearly that for agriculture to maintain both high productivity and ever-improving impacts on natural systems (both few negatives and more positives), then farmers must lead the redesign revolution. Some of this will be forced adaptation, driven by climate changes, pest, disease and weed evolution, and technological disrup-tors of economies. Some will be deliberate and chosen. As we have indicated elsewhere in this book, some of the best examples and greatest advances have been occurring in developing countries. Agricultural knowledge economies are beginning to emerge.

7

REDESIGNING AN AGRICULTURAL KNOWLEDGE ECONOMY WITH SOCIAL CAPITAL

Working together to grow natural capital

We have shown in the previous two chapters that a great many encouraging sustainable intensification innovations have emerged in both developing and industrialised countries, and on small to large farms. Central to all are farmers, their families and their communities. In many cases, the shift from efficiency and substitution to redesign can only occur if transitions are made across whole landscapes. Save for the largest of farms and rangeland properties, individual decision-making for farming occurs at a lower system level than many environmental services and assets operate. A core principle of SI centres on the positive contributions that food and fibre production activities can make to natural capital of soil, water, biodiversity and atmospheric quality. In most social systems, this implies the need for collaboration and cooperation between farmers. One farmer can change the world through leadership and demonstration; it takes many farmers in a particular landscape to build assets that will benefit both them and the planet.

As we showed in Chapter 2, many industrialised farm landscapes have lost large numbers of farmers and farm workers in recent decades – many millions in some countries. The challenge to foster collaboration in landscapes where farms are large and farmers few is greater than in contexts where farms are small and farmers many. Wes Jackson set up the Land Institute in Salinas, Kansas, to create a new vision for the emptying prairies. They have a presence in the small town of Matfield Green, which is, he wrote, "typical of countless towns throughout the Midwest and Great Plains. People have left, people are leaving, buildings are falling down or burning down. Fourteen of the houses here that do still have people have only one person, usually a widow or widower". The aim of the Land Institute is to find ways to let people become native to their places, and in the Midwest this means recreating the tall grass prairie in places and breathing life into rural towns.

This will need stepwise changes in thinking, as past incrementalism has only worked to remove people from the land. Sitting in a 75-year-old house, "abandoned more times than anyone can recall", Jackson described the decline of the town's life:

> I can see the abandoned lumberyard across the street next to the abandoned hardware store. Out another window is the back of the old creamery that now stores junk ... from a different window, I can see the bank, which closed in 1929 and paid off ten cents to the dollar ... Around the corner is an abandoned service station. There were once four! Across the street is the former barber shop.

He wrote, "this story can be repeated thousands of times across our land".

We shall see in this chapter that some of the most astonishing advances in the creation of social capital have occurred in developing countries. These have occurred in the context of irrigation management, integrated pest management, watershed and catchment management, joint forest management (JFM), farmer experimental groups, micro-credit groups. In industrialised countries, some of the best collaborations have arisen from Landcare and farmer-led watershed programmes, from farmer experimenting groups, on some individual farms and in legacy communities where a culture of collaboration has always been strong.

What is social capital?

The term social capital is now commonly used to describe the importance of social bonds, trust, reciprocity, norms and collective action. Its value was identified by Ferdinand Tönnies and Petr Kropotkin in the late nineteenth century, shaped by Jane Jacobs and Pierre Bourdieu 70–80 years later and given novel frameworks by sociologist James Coleman and political scientist Robert Putnam in the 1980s and 1990s. Coleman (1988) described it as "the structure of relations between actors and among actors" that encourages productive activities. These aspects of social structure and organisation act as resources for individuals to use to realise their personal and community interests. As social capital lowers the costs of working together, it facilitates cooperation. People have the confidence to invest in collective activities, knowing that others will do so too. They are also less likely to engage in unfettered private actions that result in resource degradation.

There are four central features of social capital: relations of trust; reciprocity and exchange; common rules, norms and sanctions; and connectedness, networks and groups. Trust lubricates cooperation, and so reduces the transaction costs between people. Instead of having to invest in monitoring others, individuals are able to trust them to act as expected. This saves money and time. It also creates a social obligation, as trusting someone engenders reciprocal trust. There are different types of trust: the trust we have in individuals whom we know; and the trust we have in those we do not know, but which arises because of our confidence in a known social structure. Trust takes time to build, but is easily damaged, and when a society is pervaded by distrust, cooperative arrangements are very unlikely to emerge or persist.

Reciprocity and regular exchanges increase trust, and so are also important for social capital. Reciprocity comes in two forms: specific reciprocity is the simultaneous exchanges of items of roughly equal value, while diffuse reciprocity refers to a continuing relationship of exchange that at any given time may be unrequited, but which over time is repaid and balanced. Again, both contribute to the development of long-term obligations between people, an important part of achieving positive sum gains for natural capital.

Common rules, norms and sanctions are the mutually agreed or handed-down norms of behaviour that place group interests above those of individuals. They give individuals the confidence to invest in collective or group activities, knowing that others will do so too. Individuals can take responsibility and ensure their rights are not infringed. Mutually agreed sanctions ensure that those who break the rules know they will be punished. These rules of the game, sometimes called the internal morality of a social system or the cement of society, reflect the degree to which individuals agree to mediate their own behaviour. Formal rules are those set out by authorities, such as laws and regulations, while informal ones shape our everyday actions. Norms, by contrast, indicate how we should act (when driving, norms determine when we let other drivers in the traffic queue; rules tell us on which side of the road to drive).

Connectedness, networks, and groups are the fourth key feature of social capital. Connections are manifested in many different ways, such as trading of goods, exchange of information, mutual help, provision of loans and common celebrations and rituals. They may be one-way or two-way, and may be long-established, and so not respond to current conditions, or subject to regular update. Connectedness is institutionalised in different types of groups at the local level, from guilds and mutual aid societies to sports clubs and credit groups, from forest, fishery or pest management groups to literary societies and mothers' groups. High social capital also implies a likelihood of multiple memberships of organisations and good links between groups.

But first, an acknowledgement of danger: it is easy to appear too optimistic about local groups and their capacity to deliver economic and environmental benefits for all. There are always divisions and differences within and between groups and communities, and conflicts can result in environmental damage. Not all forms of social relations are necessarily good for everyone in a community. A society may be well organised, have strong institutions with embedded reciprocal mechanisms, but be based not on trust but on fear and power, such as in feudal, hierarchical, racist and unjust societies. Formal rules and norms can also trap people within harmful social or family arrangements. Again a system may appear to have high levels of social assets, with strong families and religious groups, but contain abused individuals or those in conditions of slavery or other forms of exploitation. Some associations can also act as obstacles to the emergence of sustainable intensification, encouraging conformity, perpetuating adversity and inequity, and allowing some individuals to force others to act in ways that suit only themselves. At the same time, external organisations often operate cynically to serve their own interests by preventing actions that could improve community well-being, such as the merchants of doubt in the tobacco industry (Orestes and Conway, 2011).

For farmers to invest in collective action and social relations, they must be convinced that the benefits derived from joint approaches will be greater than those from going it alone. External agencies, by contrast, must be convinced that the required investment of resources to help develop social and human capital, through participatory approaches or adult education, will produce sufficient benefits to exceed the costs. Elinor Ostrom (1990) put it this way: "participating in solving collective-action problems is a costly and time consuming process. Enhancing the capabilities of local, public entrepreneurs is an investment activity that needs to be carried out over a long-term period". For initiatives to persist, the benefits must exceed both these costs and those imposed by free-riders.

One mechanism to develop the stability of social connectedness is for groups to work together by federating to influence district, regional or even national bodies. This can open up economies of scale to bring even greater economic and ecological benefits. The emergence of such federated groups with strong leadership also makes it easier for government and non-governmental organisations to develop direct links with poor and formerly excluded groups, though if these groups were dominated by the wealthy, the opposite would be true. This can result in greater empowerment of poor households, as they better draw on public services. Such interconnectedness between groups is more likely to lead to improvements in natural resources than regulatory schemes alone.

But this raises further questions. What will happen to state–community relations when social capital in the form of local associations and their federated bodies spreads to large numbers of people? Will states seek to colonise these groups, or will new broad-based forms of democratic governance emerge? Important questions also relate to the groups themselves. Good programmes may falter if individuals start to burn out, feeling that investments in social capital are no longer paying. There are also worries that the establishment of new community institutions may not always benefit the poor. There are signs that they can all too easily become a new rhetoric too, but this is an inevitable part of any transformation process. The old guard adopts the new language, implies they were doing it all the time, and the momentum for change slows. But this is not a reason for abandoning the new. Just because some groups are captured by the wealthy, or are run by government staff with little real local participation, does not mean that all are fatally flawed. What it does show clearly is that the critical frontiers are inside us. Transformations must occur in the way we all think if there are to be real and large-scale transformations in the land and the lives of people, particularly those in the drylands who have struggled with the uncertainties of rainfall, and whose challenges may be even greater in the immediate years to come.

Participation and social learning

The term participation is now part of the normal language of most development and conservation agencies. This has created many paradoxes, as it is easy to misinterpret the term. In conventional development, participation has instrumentally centred on encouraging local people to contribute their labour in return for food, cash or materials. But material incentives distort perceptions, create dependencies and give the misleading impression that local people are supportive of externally driven initiatives. When little effort is made to build local interests and capacity, then people have no stake in maintaining structures or practices once the flow of incentives stops. If people do not cross a cognitive frontier, then there will be no redesign.

The dilemma for political and social authorities is they both need and fear people's participation. They need people's agreement and support, but they fear that wider and open-ended involvement is less controllable. But if this fear permits only stage-managed forms of participation, then distrust and greater alienation are the most likely outcomes. Participation can mean finding something out and proceeding as originally planned. Alternatively, it can mean developing processes of collective learning that change the way people think and act. The ways that organisations interpret and use the term participation range from passive participation, where people are told what is to happen and tend to act out predetermined roles, to self-mobilisation, where people take initiative independently of external institutions (Pretty, 1995).

Agricultural development has often started with the notion that there are technologies that work, and so it is just a matter of inducing or persuading farmers to adopt them. But the problem is that the imposed models can look good at first, and then fade away. Alley cropping, an agroforestry system comprising rows of nitrogen-fixing trees or bushes separated by rows of cereals, was long a focus of research. Many productive and sustainable systems, needing few or no external inputs, were developed. They stop erosion, produce food and wood, and can be cropped over long periods. But the problem was that very few farmers then adopted these systems as designed. They appeared to have been produced suitable largely for research stations with their plentiful supplies of labour and resources, and standardised soil conditions.

As we have shown with Stuart Hill's Efficiency–Substitution–Redesign framework, it is critical that sustainable intensification does not prescribe concretely defined end points for technologies and practices. This only serves to restrict the adaptive options of farmers and rural people. As ecological, climate and economic conditions change, and as knowledge adapts too, so must the capacity of farmers and communities be enhanced to allow them to drive transitions. Agricultural sustainability does not imply simple models or packages to be imposed. Rather it should be seen as a process of social learning. This centres on building the capacity of farmers and their communities to learn about the complex ecological and biophysical complexity in their fields and farms, and then to act on this information. The process of learning, if it is socially embedded and jointly engaged upon, provokes changes in behaviour and can bring forth a new world.

We could think of natural and agroecosystems as being full of megabytes of information, and so ensure a focus on developing the proper operating systems for a new sustainability science. Genetics, pest–predator relationships, moisture and plants, soil health, and the chemical and physical relationships between plants and animals are subject to manipulation, and farmers who understand some of this information, and who are confident about experimentation, have the components of an advanced operating system. This is social learning: a process that fosters innovation and adaptation of technologies embedded in individual and social transformation. Thus most social learning is not to do with hard information technology, such as computers or the Internet. Rather, it is associated, when it works well, with farmer participation, rapid exchange and transfer of information when trust is good, better understanding of agroecological relationships and farmers experimenting in groups. Large numbers of groups work in the same way as parallel processors, the most advanced forms of computation.

We will see that social capital lowers the costs of working together. People have the confidence to invest in collective activities, and are less likely to engage in unfettered private actions with negative outcomes, such as resource degradation. Collective resource management programmes that seek to build trust, develop new norms and help form groups have become increasingly effective, and such programmes are variously described by the terms community-, participatory-, joint-, decentralised- and co-management.

What about agricultural extension?

This is not the place to offer a comprehensive analysis of past agricultural and rural extension, save to note that there has been considerable conceptual and practical movement away from the once dominant transfer-of-technology model. Novel technologies and practices are developed by experts, assumed to work on all farms within a particular category (for example, an agroecological zone, region or crop type) and then pushed out to farmers. Those declining to adopt are then labelled laggards or non-adopters: the fault is theirs. Agricultural extensionists have always been the lowest class of professional in agricultural development, even though they tend to work closest with farmers (Chambers, 1983). Nonetheless, almost all countries worldwide still have publically funded extension systems, in which there is recognition that public goods can be created in food systems and natural environments through the social capital created by farmers working together, and many public and private partnerships have been created.

The UK remains almost unique in having begun to close its public extension service, ADAS, from the 1980s. The space has been partly filled by conservation-based NGOs and charities, and by farm-based groups such as LEAF. Most farmers, however, now receive their advice and knowledge from buyers, such as supermarkets, or input-suppliers. This hinders transitions towards redesign for sustainability.

The loss of public extension has had a dramatic though largely hidden effect on UK farming. Five years of work with 31 farmers managing 12,000 hectares in mid-Norfolk found that there were close links between the current social capital of farmers and the sustainability of their operations (Figure 7.1). Irrespective of farm size, type of operation, soil type and degree of formal adherence to external standards, farmers with greater bonding, bridging and linking social capital were able to produce more sustainable farms (Hall and Pretty, 2008). Jilly Hall also documented the effect of the distancing of government agencies from farmers as a result of the full privatisation of ADAS in 1997, whereby all services were chargeable to farmers.

The Norfolk Arable Land Management Initiative (NALMI) farmers, particularly over the age of 45, described with warmth and animation personal narratives of their close working relationships with staff from the public extension service and Ministry officials. Their response to open questions about the nature of their relationship with these agencies was an extensive display of vividly recalled memories. Information regarding their emotional responses to the relationship was particularly forthcoming: 97 per cent were positive responses with only 3 per cent negative. But this shifted dramatically when comments were made about the current Ministry and agency officials: 87 per cent were negative, 13 per cent positive. Farmers reported that changes had created new feelings of distance between themselves and government agency staff. The degree to which they felt this distance depended largely on their age: older farmers all described more dramatic

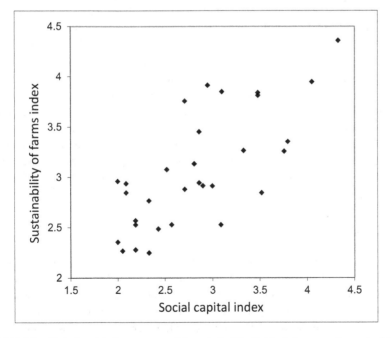

FIGURE 7.1   Relationship between social capital of individual farms and sustainability, 31 farmers, Norfolk, UK

Source: Hall (2008).

feelings of distancing. It was clear that this general decline in trust led to defensive relationships which caused a delay in farmers' transition to more sustainable land management.

The contrast with the USA, where extension is still supported by state and national agencies, is considerable. One example is work by the Kansas Rural Center, which supports family farming and the grassroots involvement of local people in farming and countryside matters. Their Heartland Sustainable Agriculture Network brings farmers together to enhance experimentation, exchange and education. The network organises farmers in small clusters to work together on issues important to them. These include Covered Acres – farmers in central Kansas experimenting with legume cover crops; Smoking Hills – farmers working on grazing management in Saline County; Resourceful Farmers – crop, livestock and dairy farmers in south-central Kansas who give on-farm demonstrations of rotational grazing and clean-water practices; and Quality Wheat – organic farmers in west Kansas seeking to improve soil fertility and increase protein content of wheat. The network is a clearing house for ideas on sustainable agriculture, helps to build support for new ideas, nurtures leadership, creates confidence amongst farmers to try something new and works with conventional agricultural institutions to build support for rural regeneration through sustainable agriculture.

Social capital is thus an important prerequisite to the adoption of sustainable behaviours and technologies over whole landscapes. Three types of social capital are commonly identified: the ability to work positively with those closest who share similar values (bonding social capital); working effectively with those who have dissimilar values and goals is bridging social capital; and the ability to engage positively with those in authority either to influence their policies or to garner resources is termed linking social capital (Hall and Pretty, 2008). Linking social capital encompasses the skills, confidence and relationships that farmers employ to create and sustain rewarding relationships with staff from government agencies. To gain the most from social capital, individuals and communities require a balanced mixture of bonding, bridging and linking relationships. This is the knowledge intensification needed for advances towards sustainable intensification (Buckwell *et al.*, 2014).

Farmer field schools

A remarkable social innovation arose in the 1980s when Peter Kenmore, Kevin Gallagher, Russ Dilts, Lou Setti, David Kuhler and Dada Morula Abubakr incorporated adult education strategies with agroecology, drawing on Paulo Freire, the civil rights movement, the approaches to education developed by the Highlander Folk and Research Group, and Danish Folk High Schools. They first showed that in irrigated rice systems the more pesticide used, the greater the pest damage. They realised that insecticides were killing beneficial insects and arthropods. They also realised that most farmers would not know this: detailed entomological knowledge is rarely a feature of local indigenous knowledge systems. Their first hypothesis: could irrigated rice management be amended to reduce insecticide use, and could the beneficials do the work of pest management. Their second: could they create a system of learning to allow farmers to demonstrate to themselves that this really works – that they would, in short, not lose all their crops and starve. The first farmer field schools were established in Indonesia and the Philippines, and in a generation have spread through many countries and benefited large numbers of farmers.

The aims of Farmer Field Schools (FFS), often called schools without walls, are education, co-learning and experiential learning so that farmers' expertise is improved to provide resilience to current and future challenges in agriculture (Table 7.1). FFS are not just an extension method: they increase knowledge of agroecology, problem-solving skills, group building and political strength. FFS have also been recently complemented by modern methods of extension involving video, radio, market stalls, pop-ups and songs (Bentley, 2009). These can be particularly effective where there are simple messages or heuristics that research

TABLE 7.1 The principal elements of farmer field schools (FFS)

- Each FFS consists of a group of 25–30 farmers, working in small subgroups of about five each. The training is field-based and season-long, usually meeting once per week.
- The season starts and ends with a ballot box pre-test and post-test respectively to assess trainees' progress.
- Each FFS has one training field, divided into two parts; one IPM-managed (management decisions decided on by the group, not a fixed formula), the other with a conventional treatment regime, either as recommended by the agricultural extension service or through consensus of what farmers feel to be the usual practice for their area.
- In the mornings, the trainees go into the field in groups of five to make careful observations on growing stage and condition of crop plants, weather, pests and beneficial insects, diseases, soil and water conditions. Interesting specimens are collected, put into plastic bags and brought back for identification and further observation.
- On returning from the field to the meeting site (usually near the field, under a tree or other shelter), drawings are made of the crop plant which depict plant condition, pests and natural enemies, weeds, water and anything else worth noting. A conclusion about the status of the crop and possible management interventions is drawn by each subgroup and written down under the drawing (agroecosystem analysis).
- Each subgroup presents its results and conclusions for discussion to the entire group. As well as in the preceding field observations, the trainers remain as much as possible in the background, avoiding lecturing, not answering questions directly, but stimulating farmers to think for themselves.
- Special subjects are introduced in the training, including maintenance of 'insect zoos' where observations are made on pests, beneficial insects, and their interactions. Other subjects include leaf removal experiments to assess pest compensatory abilities, life cycles of pests and diseases (and in recent years the expansion of topics away from just IPM).
- Socio-dynamic exercises serve to strengthen group bonding in the interest of post-FFS farmer-to-farmer dissemination.

Source: Settle and Hama Garba (2011).

has shown will be effective if adopted. Farmer field schools have now been used for soil management, biodiversity, livestock, and are known to have benefited 12 million farmers.

How effective have farmer field schools been? In Sri Lanka, 610 FFS were conducted over 1995–2002 on farms of mean size of 0.9 ha, and on which paddy rice yields improved slightly from 3.8 to 4.1 t/ha while insecticide applications fell from 3.8 to 1.5 per season (Tripp *et al.*, 2005). More than a third of farmers eliminated pesticide use completely, and an average farmer could name four natural enemies compared with 1.5 by those who had not attended a FFS. In Burkina Faso, Benin and Mali, 116,000 farmers have been trained in 3,500 FFS for vegetables, rice and cotton (Settle and Hama Garba, 2011). Pesticide use has been cut to 8 per cent (previously 19 compounds were found in the Senegal River, 40 per cent of which exceeded maximum tolerances by 100-fold), biopesticides and neem use has increase by 70–80 per cent, and there have been substantial increases in yields (e.g. rice in Benin up from 2.3 to 5 t/ha). In Mali, cotton farmers participating in FFS reduced pesticide use by just over 90 per cent compared with pre-FFS use and a control group (Settle *et al.*, 2014). In Lao PDR, 1,500 farmers on 2,200 hectares of land increased yields by 46 per cent from 2.9 to 4.25 tonnes per hectare (Ketelaar *et al.*, 2018).

FFS and IPM have also had an impact at macro level in some countries (Ketelaar and Abubakar, 2012). Between 1994 and 2007, rice farmers in the Philippines reduced pesticide application frequency and applications per hectare by 70 per cent, increased yields by 12 per cent and increased the inter-year stability of yields. Over this period, national rice production rose from 10.5 to 16.8 Mt (FAO, 2014a). In Bangladesh, pesticide use has increased substantially in recent years, and in some districts (e.g. Natore) results in 40–50 applications per season for beans, and 150–200 times on brinjal, sometimes daily (Bentley, 2009). Many farmers spray only crops for market, and keep those for home consumption unsprayed. In other regions of Bangladesh there are severe shortages of rural labour where the burgeoning garment industry has attracted young people, and here FFS have helped farmers adopt and increase herbicide use in rice whilst ensuring that fish and frogs are protected.

Evaluation of two IPM–FFS programmes in Sichuan, China, found that yields slightly increased whilst pesticide applications fell 40–50 per cent (Mangan and Mangan, 1998). In the wet rice agroecosystems, 64 per cent of the insects and spiders were found to be predators and parasites, 19 per cent neutral detritivores and only 17 per cent rice pests. Beneficials were extremely effective at controlling pests, until pesticides were used. Some 100,000 farmers have now been trained in FFS and community IPM in China (Yang *et al.*, 2014). In Cambodia, 270 farmer field schools produced a range of innovations to increase both wet and dry season rice yields (Chhay *et al.*, 2017).

Nonetheless, it is difficult to overcome the fears that farmers have; often these have been encouraged by the pesticide industry. Farmers need to overcome fears that insects always cause harm, that insect pests will transfer from sprayed to unsprayed fields and farms, and fears of crop loss (Palis, 2006). In some cases, farmers have experienced anxieties about maintaining good social relations, spraying their crops secretly at night. Yet where farmers did join FFS, they were able to have the confidence to make dramatic reductions in pesticide applications from 1.9 to 0.3 per season (van den Berg and Jiggins, 2007). Nonetheless, many illegal and toxic pesticide compounds are still being used, even where there has been great success with both FFS and IPM. In Vietnam, for example, the most toxic pesticides have been banned, but many are still getting through to farmers (Hoi *et al.*, 2016).

Further innovations in process have occurred in Uganda with the development of Agro-Pastoral Field Schools (APFS), with the training of a large pool of facilitators, trainers and implementation NGOs (FAO, 2013c). The primary aim has been to build resilience for communities subject to recurrent dryland hazards such as drought, flood and animal disease, some of these accentuated by climate change. Some 4,400 APFS have been held to 2014, attended by 117,000 agro-pastoralists, with the training of 850 facilitators and master trainers. APs build their livelihood resilience by increasing the number of intervention options they have at hand, including pest and disease management, tree nurseries, watershed management, group marketing, vegetable production, improved seeds and livestock nutrition. Each trained group develops their own collective action plan, many also creating resource-sharing plans across groups.

FFS have become so successful that it is almost impossible to know the aggregated impact. The approach has now been deployed in 90 countries, including in central and east Europe, the USA and Denmark (Braun and Duveskog, 2009; FAO, 2016d). The numbers of farmers trained in each can be in the hundreds of thousands. In Asia, 650,000 farmers have been trained in Bangladesh, 250,000 in India, 930,000 in Vietnam, 1.1 million in Indonesia, 500,000 in the Philippines and 90,000 in Cambodia. A further 20,000 FFS graduates are running FFS for other farmers, having transitioned from farmer to expert trainer. In the West African countries of Benin, Burkina Faso, Guinea, Mali, Mauritania, Niger and Senegal, some 160,000 farmers have been trained in Integrated Pest and Plant Management. Worldwide, some 12 million farmers have graduated from farmer field schools (FAO, 2016d).

The package of social engagement to spread push–pull

In order to enhance scaling up of push–pull technology (Chapter 5) to millions of farmers in Africa, innovative communication and institutional arrangements have been deployed to transfer the technology through strategic partnerships catering for different types of smallholder users. Different farmer learning approaches have been field-tested. Novel communication strategies and cost-effective scaling-up models have used to achieve faster learning and adoption of the technology, leading to food security, nutrition and environmental impacts at scale. Ranked in terms of their increasing resource- and knowledge-intensiveness from our previous research (Khan *et al.* 2017), these include print materials with local stockists, field days (FD), farmer teachers (FT) and farmer field schools (FFS). Given the nature of push–pull technology as a knowledge-intensive technology, these pathways have been used either singly or in combination to enhance the quality of information received. The following delivery mechanisms have been used:

i   Print material (PM) gives the learner clear and effective understanding of the technology functioning, components and recommended agronomic practices. PM is used in creating awareness and stimulating farmers' interest in the technology. Once stimulated, farmers may gain enough interest to seek additional information. PM has the added advantage of preserving the information and can therefore be used for a long time as a permanent reminder.

ii   Farmer teachers (FT) also referred to in the innovative farmer approach involves selection of progressive early adopters of new technologies. The method capitalises on local social networks. The concept of farmer trainers is based on the hypothesis that there are always farmers who have above average skills, knowledge and talents in farm management. These farmers motivate other farmers, help them to improve their skills, share their know-how and are therefore trained to train other farmers. Farmer trainers are identified based on their knowledge and understanding of the technology, and motivated by the recognition they receive in society. Their training is typically hands-on, initially at the trainers' farm and later at the trainees' farm with monitoring and evaluation visits by the trainer to ensure quality.

iii   Field days (FD) are day-long events commonly used in rural agricultural extension. Interested farmers are invited to a particular field or plot and specific information about the technology is demonstrated and discussed. A FD ranges from structured presentations to more informal events where participants walk through the field plot at their own pace to view the demonstrations. Farmers are able to interact with the facilitators as well as with other farmers and exchange ideas and experiences. Hands-on training and physical participation of the farmers is usually encouraged. In this approach, farmers are actively engaged in planning and training activities. At the end of a FD participants are interviewed to assess the effectiveness of FD as a knowledge transfer model.

iv   Farmer Field Schools (FFS) started as "schools without walls" where farmers take charge of the learning process by organising experiments, leading discussions, making plans and accomplishing the tasks that were initially thought to be too complex. As a group-based approach, FFS appear to offer greater learning opportunities for resource-poor or marginalised members, such as women or youth, than less interactive or supported approaches (Ramisch *et al.*, 2006; Phillips *et al.*, 2014). FFS recognise the need to involve farmers in the technology development and transfer. Farmers experiment and solve their problems independently, and are able to adapt the technologies to their own specific environmental and cultural needs. Participants are encouraged to share their knowledge with other farmers, and trained to teach the courses themselves, thus reducing the need for external support.

v   New media: innovative approaches have been integrated in the technology dissemination package, including the use of participatory video, cartoon books and drama where farmers share own experiences in local contexts, embedded within strategic partnership platforms. Audio-visual methods are more effective than printed material in disseminating knowledge-intensive technologies to farmers with low literacy levels. Access to video documentation from other farmers (participatory video) also enables smallholder farmers to learn more easily, retain knowledge about, adopt and apply push–pull to their cereal farming systems due to own farmer narrative. Moreover, farmers more readily share knowledge and their own findings if they use participatory video as part of their work (Ongachi *et al.*, 2017). Participatory communication functions as a catalyst for action and as a facilitator of knowledge acquisitions and knowledge sharing among people.

Joint and collective management for sustainable intensification

We now turn to four areas where farmer groups have been the mechanism to deliver both natural capital and individual farmer benefits. For as long as people have been engaged in agriculture, farming has been at least a partially collective business. Farmers have worked together as well as in competition, and sometimes have formally organised into named groups and structures to sustain activities over time, generations too.

This emergence of social capital manifested in groups and associations worldwide is encouraging. It is helping to transform some natural resource sectors, such as forest management, with 25,000 forest protection committees in India, or participatory irrigation in Sri Lanka with 33,000 groups. Some countries or regions are being transformed: nearly 2 million Asian farmers engaged in sustainable rice management.

However, the fact that groups have been established does not guarantee that resources will continue to be managed sustainably or equitably. What happens over time? How do these groups change, and which will survive or become extinct? Some will become highly effective, growing and diversifying their activities, whilst others will struggle on in name only. Can we say anything about the conditions that are likely to promote resilience and persistence? There is surprisingly little empirical evidence about the differing performances of groups, particularly over time. This is partly because there has remained a relative myopia in evaluations about quantifiable impacts, particularly yields, incomes and costs, relative to more qualitative indicators of participation, social networking and the development of local institutions over time.

Theoretical models have been developed to describe changes in social and organisational structures, commonly characterising structure and performance according to phases. Some of these focus on organisational development of business or corporate enterprises, with a particularly strong emphasis on the life cycles of groups. Others focus on the phases of learning, knowing and worldviews through which we as individuals progress over time (Pretty and Frank, 2000).

When groups form, they do so to achieve a desired outcome. This is likely to be in reaction to a threat or crisis, or as a result of the prompting of an external agency. They tend at this stage to be looking back, trying to make sense of what has happened. There is some recognition that the group has value, but rules and norms tend to be externally imposed or borrowed. Individuals are still looking for external solutions, and so tend to be dependent on external facilitators. There is an inherent fear of change, as members would really like things to return to before the crisis arose and the need to form a group arose. For those groups concerned with the development of more sustainable technologies, the tendency at this stage is to focus on efficiency by reducing costs and damage. In agriculture, for example, this could mean the adoption of reduced-dose pesticides and targeted inputs, but not yet the use of redesign.

The second phase sees growing independence, combined with a realisation of new emerging capabilities. Individuals and groups tend to look inwards more, beginning to make sense of their new reality. Members are increasingly willing to invest their time in the group itself as trust grows. Groups at this stage begin to develop their own rules and norms, and start to look outwards. They develop horizontal links with other groups and realise that information flowing upwards and outwards to external agencies can be beneficial for the group. With the growing realisation that the group has the capacity to develop new solutions to existing problems, individuals tend to be more likely to engage in active experimentation and sharing of results. Agricultural approaches start incorporating regenerative technologies to make the best use of natural capital rather than simple eco-efficiency. Groups are now beginning to diverge and develop individual characteristics. They are more resilient, but still may eventually break down if members feel they have achieved the original aims, and do not wish to invest further time in pursuing new ones.

The final phase involves a ratchet shift for groups, with greater awareness and interdependence. They are very unlikely to unravel or, if they do, individuals have acquired new worldviews and ways of thinking that will not revert. Groups are engaged in shaping their own realities by looking forward, and the individual skills of critical reflection (how we came here) combined with abstract conceptualisation (how would we like things to be) mean that groups are now expecting change and are more dynamic. Individuals tend to be much more self-aware of the value of the group itself. They are capable of promoting spread of new technologies to other groups, and of initiating new groups themselves. They want to stay well linked to external agencies, and are sufficiently strong and resilient to resist external powers and threats. Groups are more likely to come together in apex organisations, platforms or federations, to achieve higher level aims. At this stage, agricultural systems are more likely to be redesigned according to ecological principles, no longer adopting new technologies to fit the old ways, but innovating to develop entirely new systems.

What is interesting is the spread of group action in four areas: watershed and catchment management; irrigation water users; micro-finance; and joint and participatory forest management.

Watershed and catchment management groups

Governments and non-governmental organisations have increasingly come to realise that the protection of whole watersheds or catchments cannot be achieved without the willing participation of local people. Indeed, for sustainable solutions to emerge, farmers need to be sufficiently motivated to want to use resource-conserving practices on their own farms. This in turn needs investment in participatory processes to bring people together to deliberate on common problems, and form new groups or associations capable of developing practices of common benefit. This had led to an expansion in programmes focused on micro-catchments – not whole river basins, but areas usually of no more than several hundred hectares, in which people know and trust each other. The resulting uptake has been extraordinary, with participatory watershed programmes reporting substantial yield improvements, together with substantial public benefits, including groundwater recharge, reappearance of springs, increased tree cover, microclimate change, increased common land revegetation and benefits for local economies. Some 50,000 watershed and catchment groups have been formed across Australia, Brazil, Burkina Faso, Guatemala, Honduras, India, Kenya, Niger and the USA. As with all these examples of social capital creation, there are also many political pitfalls to achieving integrated and fair watershed management: there will always be contested claims on resources (Blomquist and Schlager, 2005; Bharucha *et al.*, 2014).

BOX 7.1 WATERSHED APPROACHES IN KENYA AND BRAZIL

Catchment approach, Kenya

The Catchment Approach was adopted by the Ministry of Agriculture to concentrate resources and efforts within areas of 200–500 hectares, so that all farms could be conserved with full community participation. Small adjustments and maintenance were then carried out by the community members themselves with the support of extension agents. A catchment conservation committee of farmers is elected to be responsible for coordinating local activities. In the 1990s, 4,500 catchment committees were formed, and by the 2000s, 100,000 farms were being conserved a year. The process of implementation of the catchment approach itself varies according to local circumstances and differing interpretations of the degree of participation necessary to mobilise the catchment community. Some still feel farmers should simply be told what to do. Others do not invest enough time in developing relations of trust. But where there is genuine participation in planning and implementation, the impacts on food production, landscape diversity, groundwater levels and community well-being can be substantial.

The soil conservation programme ended in 2000, but was then broadened in scope to form a National Agriculture and Livestock Extension Programme with support from Sweden. This programme covers about 400,000 farms per year with 4,000 extension staff, and helps farmers now form common interest groups for marketing of produce. In addition, the programme promotes the principles of human rights and democracy of participation, non-discrimination, transparency and accountability. Productivity has continued to grow remarkably, and many farmers have diversified with bananas, vegetables, dairy animals, fish ponds and bee hives. Japhat Kiara says, "the programme has benefited hugely from the soil conservation programme, farmers' yields and incomes have increased, more resource-poor and vulnerable farmers are receiving farming technology information, and staff motivated".

Microbacias of Southern Brazil

One hundred years ago, 85 per cent of the three states of Paraná, Santa Catarina and Rio Grande do Sul was under forest; now it is less than a third. Small farms dominate the land, about a half being less than 10 hectares in size. But farms on slopes brought new problems. They lose water rapidly when rain falls, and soil erosion had become a serious statewide and agricultural problem by the early 1990s. It was at this time that the state government's research and extension agency, EPAGRI, began the *microbacias* programme, working in 500 catchments to encourage all farmers to adopt conservation tillage methods to cut erosion and conserve water. They helped conservation tillage to spread to more than 1 million hectares by the early 2000s. An explicitly participatory approach was critical, working closely with farmers on technology development and innovation. As a result, more than 100,000 farmers have benefited, as their yields have increased whilst input costs have fallen. Adopting zero-tillage methods requires many innovations: new ways of direct seeding or for rotary hoeing, roller-blades for cutting cover crops, mini-tractors and carts, some hand-pulled, some by animals, others motorised. A variety of farmers' associations have emerged to help in technology development and sharing, and there are 6,000 groups sharing machinery, thus enabling farmers to access technologies that would normally only be available to large operators.

Participatory irrigation management and water users' groups

Although irrigation is a vital resource for agriculture, water is rather surprisingly rarely used efficiently. Without regulation or control, water tends to be overused by those who have access to it first, resulting in shortages for tail-enders, conflicts over water allocation, and waterlogging, drainage and salinity problems. But where social capital is well developed, then water users' groups with locally developed rules and sanctions are able to make more of existing resources than individuals working alone or in competition. The resulting impacts, such as in the Philippines and Sri Lanka, typically involve increased rice yields, increased farmer contributions to design and maintenance of systems, dramatic changes in the efficiency and equity of water use, decreased breakdown of systems and reduced complaints to government departments. More than 60,000 water users' groups have been set up in the past decade or so in India, Nepal, Pakistan, the Philippines and Sri Lanka.

It has become clear that the inefficiencies of public administration produced market failures in managing irrigation water, thus demanding new principles of organisation. Participatory Irrigation Management and associated Water User Groups or Associations (WUAs) emerged as both concepts and practice that have spread substantially. In Mexico, 2 million of the 3.2 Mha of government-managed systems have been transformed by WUAs, and half the systems in Turkey have been turned over to local groups (Groenfeldt and Sun, 1997). In China, a quarter of all villages have WUAs, and these have reduced maintenance expenditure whilst improving the timeliness of water delivery and fee collection. Amongst some WUAs, farm incomes have improved whilst water use has fallen by 15–20 per cent (Wang et al., 2010). In Bali, there are 1,800 long-standing self-organising irrigation groups that cover nearly 20 per cent of the rice area (Kulkarni and Tyagi, 2012).

The Andhra Pradesh Farmer Managed Groundwater Systems Project was set up in seven districts in the state of Andhra Pradesh. Farmers have no access to surface irrigation infrastructures, nor the capacity to invest in borewells. The project thus instituted a novel arrangement whereby groups of farmers were given access to borewells collectively managed through associations. Farmers were trained in the concept of a hydrological unit, raising awareness about the water source being tapped, as well as being given information on improved water management practices. Farmers trained are now well versed in relevant concepts such as recharge rates, evaporation losses, soil moisture content, crop water requirements and appropriate water-saving cropping patterns. Farmers' associations thus formed had the added benefit of being able to use their social capital to bring in additional government support, access development schemes and develop new marketing solutions to bring crops to the market (Mittra et al., 2014).

WUAs in India cover 15 Mha, but still only 12 per cent of the irrigated area (Sinha, 2014). Many of these are thought to exist only on paper, and in some areas have been subject to variable performance, elite capture and irrigation department control (Reddy and Reddy, 2005). In some contexts, rights transfers to landowners and tenant farmers have led to landless and fisher families being losers. In all collective management, the distribution of winners and losers remains a challenge: in irrigation management this centres particularly on differential benefits for farms at the head, middle and tail of systems. Nonetheless, there is also strong evidence from India that WUAs can lead to increases in the area under irrigation (more efficient), greater equity (improved benefits for tail-enders) and greater recovery of water charges (a measure of improved yields) (Sinha, 2014). In the state of Andhra Pradesh, there are 10,750 WUAs for 1 million farmers, covering almost half of the command area under participatory irrigation. Only about half of these groups are thought to be active institutions, but in those that are active, paddy rice yields are 50–60 per cent higher. Where groups stay together, initial successes have also led to resilience over time. For example, in the state of Maharashtra, community rules on irrigation and water management mean that the villages of Ralegaon Siddhi and Hivre Bazaar stay green and productive even during dry years, when the surrounding semi-arid watersheds struggle. Where groups can function effectively, and a new identity is created – the result of crossing a cognitive frontier, collectively – results can persist over time. In Nepal, around 70 per cent of all irrigated areas are managed by farmer groups (Pradhan, 2000). In Turkey, ten years of WUAs increased cropping intensity and increased yields by 53 per cent (Uysal and Atış, 2010).

Micro-finance institutions for rural social capital

One of the great recent revolutions in developing countries has been the emergence of new credit and savings systems for poor families. These are enablers for agricultural transformations. Families lack the kinds of collateral that banks typically demand, appearing to represent too a high a risk, and so are trapped into having to rely on money-lenders who charge extortionate rates of interest. A major change in thinking and practice occurred when professionals began to realise that it was possible to provide micro-finance to poor groups, and still ensure high repayment rates. When local groups, in particular women, are trusted to manage financial resources, they can be more effective than banks. This might not seem a direct foundation for sustainable intensification of agriculture, but the outcomes bring opportunity to escape poverty for both farmers and landless, as well as the particular benefits of social organisation.

Three leading innovative institutions are from Bangladesh: the Grameen Bank, the Bangladesh Rural Advancement Committee (BRAC) and Proshika (Grameen Bank, 2017; Proshika, 2017). All form groups, all work primarily with women, and all members of groups save every week in order to create the capital for lending. Grameen has 8.9 million members in 1.38 million groups spread over 81,000 villages: 97 per cent of its members are women. In 1990, Grameen had 0.17 million members, this grew to 0.5 million by 2000. The average loan is $100, for which no written contract is made between Grameen and borrower. The system works on trust, and payback rates are at 98 per cent.

BRAC has 5.4 million members in 108,000 groups, these tending to be larger than those of Grameen. BRAC takes a deliberately integrated approach to poverty pockets, especially in wetlands, on riverine islands and for indigenous populations. Through a single platform it provides agricultural and skills support, education, legal services, health care and loans. It runs 22,000 primary schools, and a rural university. More than 130 of its women members have been elected into government structures. BRAC has also diversified into social enterprises: for rural artisans (65,000 members), providing livestock insemination services, supplying dressed chicken for retailers, cold storage for potato farmers, dairy milk processing (for 50,000 farmers), services for fish farmers, tree seedlings, handmade paper, iodised salt for rural consumption, seed services, and sericulture and silk production.

Proshika works in both rural villages and urban slums, with 2.8 million people organised into 148,000 groups. Grameen, BRAC and Proshika together have some 1.64 million groups. This represents a remarkable 17 million people organised into groups built on trust and cooperation. Micro-finance institutions are now receiving worldwide prominence: the 50 micro-finance initiatives in Nepal, India, Sri Lanka, Vietnam, China, Philippines, Fiji, Tonga, Solomon Islands, Papua New Guinea, Indonesia and Malaysia have 5 million members in 150,000 groups. The developing world does not have a monopoly on micro-finance: by the end of the 1990s, there were also 200 microcredit programmes in the USA, serving 170,000 people (Anthony, 2005).

Community-based forestry

In many countries, forests are owned and managed by the state. In some cases, people are actively excluded. In others, some groups are permitted use-rights for certain products. Governments have not been entirely successful in managing forests for both biodiversity and local use, and in recent years have begun to recognise that they cannot hope to protect forests without the voluntary engagement of local communities (Blomley, 2013; Nightingale and Sharma, 2014; Liu and Innes, 2015; FAO, 2016f). There are a wide range of terms for schemes across all continents, including from community-based forestry (CBF), participatory conservation, social forestry, forest farmer cooperatives, forest protection committees and forest user groups. These CBF initiatives and institutions increase the role of local people in governing and managing forest resources, and include customary and indigenous practices as well as government-led initiatives. It is clear that CBF has close links to the sustainable intensification of agriculture, as groups with strong social capital often work on a number of land management issues together, and as forest resources supplement food and other resources needed at household level. As sustainable intensification explicitly seeks to expand productivity on existing agricultural land, it by definition implies opportunities for addressing and managing non-agricultural land for better outcomes, such as for biodiversity and timber and other products.

FAO (2016f) estimates that CBF is practised on 730 Mha by some 300,000 groups in 62 countries. This comprises 28 per cent of the world's 4 billion hectares of forest. Notable achievements have been in China (109 MHa), Australia (42 Mha for indigenous groups), Mexico, where 80 per cent of the country's forests are under community legal jurisdiction and managed by 8,000 community groups, and Vietnam, where 115,000 Forest Farmer Cooperatives have formed to manage 9 Mha of forest land. Early advances were made in India and Nepal, where experimental local initiatives in the 1980s so increased biological regeneration and income flows that governments issued new policies for joint and participatory forest management in India in 1990 and in Nepal in 1989–1993. These encouraged the involvement of non-governmental organisations as facilitators of local group formation, as governments realised they were not good at doing this themselves (Nightingale and Sharma, 2014).

Some 30,000 forest protection committees and forest users' groups were formed in India and Nepal, covering more than 25 Mha of forest, mostly with local own rules and sanctions (MFSC, 2013). In the best, benefits included increased fuelwood and fodder productivity, improved biodiversity in regenerated forests and income growth amongst poorest households. Some old attitudes have changed, as foresters came to appreciate the regeneration potential of degraded lands, and the growing satisfaction of working with, rather than against, local people (Ravindranath and Sudha, 2004). In Bangladesh, groups have helped in the shift towards upland sustainable intensification in areas formerly dominated by shifting cultivation (Nath *et al.*, 2016).

There have also been difficulties. Landmark policy changes were significant, the emergence of local groups positive, but the institutional interactions between government forest departments and communities often remain divergent (Behera and Engel, 2006; Chhatre and Agarwal, 2009). In 147 JFM villages in Andhra Pradesh, 18 per cent of villages said the forest department remained dominant, while 62 per cent said they were cooperative. Rent seeking by forest officials was still seen to be an impediment to positive outcomes on the ground. Nonetheless, the numbers in some states are significant: 2,300 forest protection committees in West Bengal protecting 320,000 ha, 40 hill resource management societies in Haryana on 15,000 ha, 400,000 hectares protected by groups in Orissa and Bihar. JFM also remains significant for the majority of India's 55 million tribal people.

The key to success in most countries is explicit government support through policies and legal protection. There have been community forest decrees in the Democratic Republic of Congo (2014), Land Use Certificates to households in Vietnam (2010), community and family forest management in Brazil (2009), devolved and participatory pilots of village control over forests in China, forest policy and legislation for JFM in Tanzania, community rights to trees in Niger and Burkina Faso (from 1985) and village-based natural resource management committees in Malawi. In China, the transfer of ownership from state to private families has resulted in 88 million households managing small forest plots (an average of 0.7 ha each) on a total of 64 Mha. Many of these households have joined together to form some 133k Forest Farmer Cooperatives. Natural capital benefits have included forest cover increases in landslide-risk slopes, patch consolidation of forests, reduced incidence of fire, reduced use of slash-and-burn agriculture, more wood volume and stems per hectare (FAO, 2016f).

Cross-country studies have illustrated common features of community natural resource management (Berkes and Ross, 2013; Baynes *et al.*, 2015): a commitment to involve community members and their existing institutions in devolving power and rights from government, a desire to integrate social and environmental objectives, a tendency to defend existing local rights and traditional values, and the use of local ecological knowledge. All of these can help build social and ecological resilience.

Redesign for sustainable intensification

A knowledge economy is emerging for many farmers across the world. This is a good reason for optimism around sustainable intensification and the multiple emergences of locally appropriate systems redesigned to enhance both food (and other valued agricultural produce) and natural capital. For redesign, we will need to move from single-loop learning to double and even triple. Single-loop learning is included in policy processes though formal evaluation. But double- and triple-loop learning become important for adaptive and transformative governance (Pahl-Wostl, 2009). Double-loop learning helps to question the assumptions behind the questions we ask and can thus lead to reframing, which is a fundamental process for disseminating new ideas and narratives (Argyris and Schön, 1978). Triple-loop learning is needed to trigger changes in governance structures and for transformation of the context. It reconsiders values and beliefs when assumptions no longer hold. Triple-loop learning is associated with paradigm shifts that rewrite norms and transform institutions (Armitage *et al.*, 2008; Hill, 2015). Both reflection and anticipation are needed for double- and triple-loop learning.

Interactive methods are now recognised as important knowledge communication strategies for developing capacities of resource-poor farmers (Berdegué and Escobar, 2001). They are an improvement on the traditional top-down linear dissemination approaches that had limited success in catalysing adoption of technological innovations (Röling, 1996; Gandhi *et al.*, 2009), shifting communication strategy from instrumental to more participatory and decentralised approaches (Leeuwis and Van den Ban, 2004). The new media, used in combination with social organisation and mediated learning, are complementing the traditional dissemination methods and resulting in improved knowledge acquisition, retention and correct application, cost-efficiency, local relevance and in stimulating real adoption.

A common concept is the platform: the notion of a space or place for engagement and performance. Innovation platforms in West Africa have resulted in increased yields and income for both maize and cassava systems (Jatoe *et al.*, 2015; Sanyang *et al.*, 2015). Farmer cooperatives have done the same in China (Song *et al.*, 2013), some putting agroecological and cultural objectives higher than just productivity. In Bangladesh, similar platforms have led to whole system change for rice production. Direct seeded rice and early maturing varieties have changed patterns of both wet and dry season farming, increasing incomes by $600 per hectare, and substantially reducing labour costs (Malabayabas *et al.*, 2014). Again, these emphasise the importance of recognising the different ecological and social circumstances of farmers and their preferences for different pathways towards sustainability (Chantre and Cardona, 2014). In all successful cases, there have been facilitators curating the new redesign (Sanyang *et al.*, 2016).

Another innovation recently developed and deployed in China to increase the sharing of knowledge and skills between scientists and farmers (Zhang *et al.*, 2016), Science and Technology Backyard Platforms (STBs) were established first in Quzhou County by China Agricultural University on the North China Plain. Now 71 STBs operate in 21 provinces, covering a wide range of crops: wheat, maize, rice, soybean, potato, mango, lychee, vegetables. STBs bring agricultural scientists to live in villages, and use field demonstrations, farming schools and yield contests to engage farmers in externally and locally developed innovations. Over six years, STBs have seen yields of wheat and maize increase by 8 per cent over control villages, rising to 23 per cent for the leading farm innovators (in some cases, with higher yields than research stations), while the use of externally derived nitrogen declined by 32 per cent. Some 50,000 farmers have been engaged and benefited, though the area covered is relatively small as each farm family has typically only 0.6 hectare. Reflections on success centre more on in-person communications, socio-cultural bonding and the trust developed amongst farmer groups of 30–40 individuals.

In Cuba, the Campesino-a-Campesino movement has developed an approach to agroecological integration that is redesigning systems (Rosset *et al.*, 2011). It is centred on a Freirian social communication method using radical educational principles (see Paulo Freire, 1970). Farmers spread knowledge and technologies to each other through exchanges, teaching and cooperatives. There are 100,000 peasant farmers of Campesino-a-Campesino in Cuba: the productivity of this sector has increased by 150 per cent over ten years, and pesticide use is down to 15 per cent of former levels.

Wisdom networks, north-east Thailand

The success elsewhere of the green revolution in raising rice productivity had only partially reached the Thai north-east, and when the Asian economic collapse of the late 1990s occurred, many families found themselves saddled with debt for agricultural inputs that they could not pay off. In addition, the income remittances sent by family members working in urban centres dried up, leaving rural families with no obvious options. Sawaeng Ruaysoongnern of Khon Kaen University observed, "this extreme failure caused people to think differently".

The first reaction from farmers was to step back from the monocultures that had become increasingly common since the 1960s, and begin to recreate polycultures. They diversified with vegetables, herbs, animals, fruit and multipurpose trees, and organised into groups. These soon connected into networks across the region. They developed an active recruitment system – each farmer aims to recruit two more each year, and each group of ten farmers seeks to establish another group each year. The consequences have been remarkable. Farmers have become more self-sufficient, growing more of their own food, but also earning more from better links to consumers in local and regional markets. The aim is to have a million members – this will be needed if regional landscapes and whole communities are to change for the better.

There is a great diversity in these local wisdom networks. The Organic Rice Network in Yasothon is led by Vichit Buonsoong, and has 2,000 core members and 20,000 network members. All farmers no longer use inorganic fertilisers and pesticides. A network with 760 members focuses on agroforestry – though they translate this to mean more than just trees: agroforestry is not about planting trees on agricultural land, but the development of new relationships between human and natural systems. In a drier region, another network of 170 farmers in 20 villages is working to establish trees so as to reduce water tables and reclaim saline seeps that are threatening crops. And yet another is focusing on intensive polycultures, with the aim of producing enough food (other than rice) for an entire family from one rai of land (a sixth of a hectare) – the one rai per family concept.

The average farm size for the region is less than 2 hectares, yet these very small farms have strength in numbers. These networks use modern technology to share knowledge, ideas and innovations where they can. They have created video profiles of individual farmers to distribute across the region. These help to tell a story to consumers and policymakers too. Even though these farmers' networks are beginning to influence local and national policies, many farmers themselves say happiness is the most important thing for them – not maximising production through debt accumulation, or sales of land for development speculation. This model for redesign is different and again instructive.

Agreco Farmers' Organisation, southern Brazil

Agreco is an ecological farmers' association based in the eastern hillsides of Santa Catarina that has expanded activities into ecotourism, school meals and links to wider economic development. They were set up in the mid-1990s by a dozen families near Santa Rosa de Lima, and began organic production of legumes, honey, grain and fruit. Though there are now 300 families involved in the network, Agreco does not intend to grow itself further, preferring to help others to set up in the same way. Produce is now sold directly to urban consumers in mixed baskets, to selected supermarkets and to schools for children's dinners. An innovation has been the emphasis on agri-tourism: the French *accueil paysan* approach was adopted, with accommodation in farm buildings, and families actively seeking to share life experiences with visitors. Agreco has formed two regional forums – one to link public and private agencies, which has already had success with energy supply, rural transport and school meals, and another to link urban with rural development. This alternative development model links an ecological approach to farming, organised small businesses, connections to food consumers and wider economic development. Sergio Pinheiro explains, "initial resistance may be explained because individualism is normal for most people. The ability and enthusiasm to work in groups has increased among farmers, and participation and trust have grown too". This resistance is common elsewhere, as farmers who are not organised often feel that they will lose something by collaborating. Oddly, such cooperation was fundamental to all agricultural and resource management systems throughout history.

Audit of groups formed

We draw together in Table 7.2 a summary of the numbers of groups in this wide range of programmes and projects to illustrate the depth of social capital formed, and the human capital now deployed in redesign. This is not a comprehensive analysis of all advances towards sustainable intensification: some 10,000 farmers are using management-intensive rotational grazing systems in the USA on 1.6 Mha, but there is no data on how many are organised into groups. Conservation agriculture is now being practised on 180 Mha worldwide, but again there is no data on how many are formally organised into innovation and development groups. In Vietnam, 1M farmers are using the system of rice intensification; again there is no data on group formation. In many of these cases, groups may well simply be existing cultural structures, such as village, community or multi-family groups.

This data indicates there has been a continued growth in the formation of social capital over the past 15 years, rising from some 500k groups to 2,688k (Pretty, 2003). This is a remarkable social transformation led by agriculture: more than 2.7 million groups with some 75 million members.

TABLE 7.2 Number of groups formed for a range of sustainable intensification activities

Focus of activity	Number of groups (thousands)
Famer field school IPM groups	480
Watershed and catchment groups	50
Participatory irrigation and water user groups	60
Micro-finance groups	
Bangladesh	1,636
Rest of world	150
Community-based forest groups	
China	133
Vietnam	115
Nepal	18
Rest of world	134
Farmer research platforms	
China: Science & Technology Backyards	1.4
Thailand: Farmer wisdom networks	0.6
Cuba: Campesino a Campesino	2.0
Farmer-to-consumer direct sales	7.4
USA: community-supported agriculture groups	1.6
France: AMAP direct sales groups	
Total	2,688

Redesign for the knowledge economy

Economic development is often characterised as being rather linear, with agrarian patterns replaced by manufacturing, these in turn displaced by services, from which then emerges the knowledge economy. The idea of the knowledge economy, however, was helpful when considering sustainable intensification and the future of agriculture. Growth and success in knowledge economies is seen as dependent on the generation and accessibility of information (OECD, 1996). Academic institutions and companies engaging in research and development are key foundations, as are communication and information technologies. Yet very rarely in discussions about knowledge economies does the sector of agriculture appear. Robotics and artificial intelligence will play a role on some farms, perhaps a majority eventually. But what we have described here in the context of sustainable intensification is the need for a knowledge economy in agriculture, in which all actors play a role in knowledge generation, sharing, deployment, evaluation and looped learning. In this way, continuing redesign could be an outcome. Agriculture could thus be a key component to the emergence of greener economies worldwide. These will take their cue from emerging ecocultures (Boehm *et al.*, 2014), where lifestyles converge around lower material consumption, but higher levels of personal and planetary well-being.

8

SUSTAINABLE INTENSIFICATION FOR GREENER ECONOMIES

Greener economies

Though much has been achieved towards sustainable intensification of agriculture there is as yet tremendous untapped potential. Progress has been achieved along two fronts: reducing harm, and building capitals. It is through this latter route that we see sustainable intensification as a pathway to the spread of green economies at scale. In this chapter, we outline how, and what would be required to achieve this.

Greener economies have become important targets for national and international organisations, including the OECD (2011), UNEP (2011), World Bank (2012), the Rio+20 conference (UNCSD, 2012) and the Global Green Growth Institute (2012). UNEP (2011) defines the green economy as "resulting in human well-being and social equity, while significantly reducing environmental risks and ecological scarcities". Perhaps most relevant to our vision is the concept of a *transformational* green economy. This goes beyond narrow, techno-centric or incremental change aiming to maintain the status quo. Instead, transformation is an essentially political project aiming at radical and inclusive change (Meadowcroft, 2011; Ferguson, 2015). Relevant to all sectors of economies will be important questions about material consumption and, in particular, how modes of consumption based on *enough not more* can be created, so resulting in mass behaviours of *enoughness* (Dietz and O'Neill, 2013). This is the safe and just operating space defined by Raworth (2017) in the doughnut: inside the planetary boundaries, and beyond a core of essential needs and values.

Political commitment is essential to this, but so far deep political commitment is rare. There are good reasons for this to change. In 2007, Stern pointed to the economic value of early action with respect to climate change: the cost of stabilising all greenhouse gases was a "significant but manageable" 1 per cent of global GDP, but a failure to reduce emissions would result in annual costs of 5–20 per cent of global

GDP. Sustainable food and agriculture systems generate wealth across the economy, including higher incomes for farmers, community assets in terms of healthy eco-systems and health benefits across the population.

Though visions vary, all recognise that innovation is key to the global green economy, both in terms of new technologies and practices (OECD, 2011) and in terms of deeper structural transformation (Barbier, 2015). Much innovation towards the green economy is occurring at the intersection of land management, food production and poverty alleviation. Around the world, food-based initiatives spearheaded by local communities are experimenting with radically new ways to grow, process and distribute food.

Community-scale innovations have been called by many terms – 'green shoots' (Macy, 2005), 'ecocultures' (Boehm et al., 2014) and most recently 'seeds' (Bennett et al., 2016). All centre on the proposition that currently dominant ways of organ-ising economies and societies harm both people and planet, and that change is both desirable and possible. All are attempts to prefigure social–ecological collapse, working to intentionally reduce the chances of catastrophic disintegration (Davy et al., 2017).

Many of the cases and examples we have summarised in previous chapters qual-ify as 'seeds' of radical social–ecological innovation. Community land and water management, community forestry, new community farming initiatives and ecosys-tem regeneration involve dramatically changed aspirations and consumption pat-terns, shifting consumption of both the poor and the affluent towards improved nutrition, increasing well-being and protecting natural capital. In many cases, farm-ers practising sustainable intensification have been able to have a direct impact on local economies in a variety of ways. Adding new elements – fertiliser trees, fodder shrubs, small livestock or fish – opens up new possibilities for better nutrition in the community, new income streams and new businesses supporting a new entre-preneurial ecosystem. New patterns of consumption emerge. In sub-Saharan Africa, the inclusion of small livestock or home gardens has given women and children access to nutritious food close to the household. In the Global North too, participa-tion in home gardens, community gardens and community-supported agriculture gives people access to fruits and vegetables, along with new levels of food and eco-logical literacy. In these niches, innovations in growing and redesign of agricultural systems can become the nexus around which a green economy can emerge and flourish. The resulting local green economies look very different to the status quo.

But new ways of doing things at community scale must be matched by vision and resolve across society and in politics. And here, the picture remains mixed.

Policies in some countries are actively promoting greener agendas, including in China, Denmark, Ethiopia, South Africa and South Korea. Such a pursuit of greener economies could lead to a new industrial revolution (Stern and Rydge, 2012) and promote further sustainable intensification of agriculture as part of a new relationship between societies and the land. Some countries are clear front-runners in terms of the scale of change. China has invested $100 billion since 2000 in eco-compensation schemes, mostly in forestry and watershed management. A total of

65 countries have implemented feed-in tariffs to encourage renewable energy generation (Renewables 21, 2017), and by 2016, renewable energy sources had grown to supply 19 per cent of global energy consumption and 24 per cent of electricity, the fastest growing sector being solar photovoltaic (PV). This alone could have a significant impact on remote rural communities, and thus lead to changes in agricultural and food systems.

But there is much to be done.

The revenue of many poorer countries is absorbed by the costs of oil imports: for example, Kenya, Senegal and India spend 45–50 per cent of export earnings on energy imports. Investing in renewable energy benefits these three countries by saving export earnings, increasing self-reliance and improving domestic natural capital. Kenya introduced feed-in tariffs on energy generated from wind, biomass, hydro, biogas, solar and geothermal sources from 2008 (UNEP, 2011). Incumbent actors who have a stake in the status quo can also block transitions (Fisher and Newig, 2016), either overtly refusing to support them, lobbying against change or simply by controlling the dominant narrative. A good example of the influence of powerful incumbent actors is in the sphere of pest control: public funding has been reduced on the assumption that pesticide companies should invest. But companies have little incentive to invest in research into alternatives which would reduce their sales. Similarly, in the absence of public extension services, farmers often rely on input producers for advice on how much to apply (Dudley *et al.*, 2017).

The continuing challenge for sustainable intensification

Pest management exemplifies the need for continuing active intervention for SI: the job is never done. Ecological and economic conditions change. We noted the spread of the papaya mealybug in Chapter 1. Another example is the cassava pink mealybug: first reported in the greater Mekong region of Thailand in 2008, it quickly spread to infest 200,000 hectares by 2010 (FAO, 2013d). The CM–IPM programme was developed with multiple tactics: ploughing and drying soil, soaking stalk cuttings in insecticide, burning of infested plants, no transport of infested plant materials and the release of *Anagyrus* parasitoids. In 2010–11, 6 million pairs of *Anagyrus* were released in Thailand, which brought the pest completely under control by 2013 (from 200,000 to 10 hectares). But the risk of spread to other countries in the region remains high.

Old pests also return: the brown planthopper (BPH) has now been called the "ghost of green revolutions past" (Bottrell and Schoenly, 2012). It was the primary threat to rice in the 1960s, was a primary driver towards the development of FFS at the end of the 1980s, yet resurfaced as a major pest threat in the 2000s owing to resistance to continued overuse of insecticides, the heavy use of nitrogen fertilisers, and changes to climate and host ranges. BPH is often triggered by overuse of insecticides, often then reinforcing farmers' fears of insect pests, provoking in them the wish to apply more. In China, between 6–9 Mha were infested with BPH in 2005–07, up from 2 Mha in the 1990s (Bottrell and Schoenly, 2012). Farmers in

China apply on average 180 kg N/ha to rice as fertiliser, and N-enriched plants are known to enhance size, performance and abundance of herbivorous pests.

These challenges suggest the need for agroecosystems to be redesigned as multifunctional landscapes to deliver a range of ecosystem services, including food production, but also water and soil conservation, soil carbon storage, nutrient recycling and pest control, all of which underpin and impact on food production. Hartley (2017) has indicated that sustainable intensification cannot be considered as shorthand for increasing agronomic efficiency; we need a broader view of agroecosystems to deliver agricultural management practices which are truly sustainable. Recently Gunton et al. (2016) proposed the following approach: "Sustainable intensification means changes to a farming system that will maintain or enhance specified kinds of agricultural provisioning while enhancing or maintaining the delivery of a specified range of other ecosystem services measured over a specified area and specified time frame." This encapsulates the importance of appropriate temporal and spatial scales, and emphasises the need to redress the balance between *sustainability* and *intensification* by moving non-production ecosystems services centre stage.

Delivering this sort of rebalancing of the agenda away from intensification and towards agroecological approaches will require innovative thinking, cooperation between disciplines and a willingness to be open to radical solutions by engaging effectively with a range of stakeholders beyond the usual suspects. For genuine sustainability, which harnesses the benefits of ecological approaches effectively, we need systems that understand or, even better, transform the relationships between the social, the ecological and the technological aspects of agriculture.

Policies and incentives to adopt sustainable intensification

Enabling policy environments are crucial for the adoption of agricultural systems that deliver both public goods (natural capital) alongside private (increased food and fibre) over time. Policy intervention in agricultural systems has clearly worked to increase output, such as during the Asian green revolutions, but intensification may involve trade-offs between provisioning ecosystem services (food production) and regulating and supporting services (Firbank et al., 2011). The key question is: can it also address challenges such as improving natural capital, nutritional security and social–ecological resilience? Global-scale policy leaders are increasingly focused on these wider goals. Recently, the FAO made the case that agricultural policies need to emphasise nutrition, and can improve nutritional outcomes by emphasising research and development that is inclusive of smallholders, focusing on important non-staple, but nutritionally dense, foods and integrated production systems (FAO, 2013e). Similarly, there is an effort to spread awareness of climate-smart agriculture (FAO, 2013b) and save and grow models (FAO, 2011) that build natural capital while improving yields and nurturing resilience.

At regional and national scales, notable successes include the combined spread of nitrogen fertiliser use and agroforestry systems in Malawi, land reform in China in the 1980s, common agricultural policy (CAP) reforms in the European Union to emphasise payments for environmental services, and tree use regulations in the Sahel. Yet, challenges remain. Sustainable intensification represents a suite of approaches, methods and technologies that can deliver more food per area of land and improve natural capital. However, most national and international policy environments are still configured to favour food production; some still actively result in damage to natural capital.

One set of policy options centres on the principle of payments for ecosystem services, whereby farmers are paid for their contributions to defined services with monetary value. Few examples have yet to work, even where there is research to show how payment for ecosystem service could work, such as Kenya's Amboseli Park with regard to compensating farmers for not cropping lands that cross elephant migration routes (Bulte *et al.*, 2008). Other options include market chain development, though usually this means having a subset of consumers willing to pay price premiums for certain products.

In general, policymakers and regulators have found it easier to seek to prevent practices or problems, such as the regulation of certain pesticide compounds, or the establishment of safe drinking water limits for certain compounds. Some have called for a revision of the regulatory frameworks governing new crop varieties, with a view to realising potential benefits for more food crops. Federoff (2010) called for an authoritative assessment of existing data on GM safety, including protein safety, gene stability, toxicity, nutritional value, allergenicity, gene flow and impacts on other organisms. This, it was conjectured, would reduce the complexity of the regulatory process.

It has been harder to encourage positive practices. This is where the concept and practices of sustainable intensification offer opportunities to engage with the wider challenges of agricultural production in a sustainability context. Norse (2012) highlights the need for creating an evidence base to support decision-making on low-carbon agriculture, and also mentions the need to create awareness of existing approaches which can entail "a win–win–win change that can be justified in terms of short-term economic, social and environmental benefits" in addition to longer term social–ecological benefits.

The US National Research Council (2010) stated that "sustainability is best evaluated not as a particular end state, but rather a process that moves farming systems along a trajectory towards greater sustainability". This suggests that no single policy instrument, research output or institutional configuration will work to maximise sustainability and productivity over spatially variable conditions and over time. The NRC (2010) made a series of recommendations regarding public research for public goods, and integration of agencies to address multidisciplinary challenges in agriculture. It was recommended that the national (in this case, USDA) and state agricultural institutions should continue publicly funded research and development of key farming practices for improving sustainability and productivity, and that federal and state agricultural research and development programmes should deliberately pursue integrated research and extension on farming systems, with a focus on whole agroecosystems.

It was further suggested that all agricultural and environmental agencies, universities and farmer-led organisations should develop a long-term research and extension initiative to understand and shape the aggregate effect of farming at landscape scale. Researchers were encouraged to adopt farmer-participatory research and farmer-managed trials as critical components of their research. At the national level, there should finally be investment in studies to understand how market structures, policies and knowledge institutions provide opportunities or barriers to expanding sustainable practices in farming. This is particularly important in enabling farmers to navigate the complex and evolving trade-offs between resource conservation and increasing farmers' incomes through participation in markets. Policies designed to conserve resources and *stabilise* resource availability and farmers' incomes may not work well over time given market imperatives to *maximise* resource use and incomes (see Bharucha *et al.*, 2014).

In the context of developing countries, 30 African cases of SI (Pretty *et al.*, 2011a, 2014) have illustrated key lessons regarding policy challenges. These projects contained many different technologies and practices, yet had similar approaches to working with farmers, involving agricultural research, building social infrastructure, working in novel partnerships and developing new private sector options. Only in some of the cases were national policies directly influential.

A clear policy need is for increased state involvement in agricultural research, development and extension. These need not entail the top-down, monolithic and centralised institutional structures which characterised previous state responses to agricultural problems. Support for these centralised structures has declined as a result of their limited success and because of the growing recognition that decentralised, adaptive and flexible institutional structures produce better outcomes. However, this has resulted in disinvestment in state support for agricultural research and development, rather than in a transformation of previous approaches. Where extension systems have been closed or underfunded, countries lack the institutions required to support farmers. Where they have been supported over the long term, significant positive gains have ensued. For example, in Ethiopia, the Ethiopian Institute of Agricultural Research (EIAR) has played a significant role in scaling up the adoption of the *Quncho* tef variety.

In part, the increased involvement of the private sector at all points in the agricultural production chain is filling this gap via a bricolage of institutions and organisations that provide different kinds of support to farmers. However, the engagement of the private sector cannot act as a substitute for state governance, or produce sustainable outcomes within a governance vacuum. The state has a clear role in shaping the environment within which an emergent private sector can contribute to sustainable intensification, through the provision of clear policy goals, holding private actors accountable to national laws and regulating their actions.

Supportive policy frameworks in developing countries

In addition to the right technologies (seeds and breeds and their agronomic–agroecological management) and social infrastructure, ideally policy environments would be supportive of sustainable intensification and its requirements. In most cases, however, agricultural policy or domestic or international policy has been generally unhelpful rather than enabling. Many successes have emerged despite policy rather than because of policy. The exceptions, however, show that activities can be greatly scaled up with the appropriate policy support.

As we saw in Box 7.1, the Kenyan National Agriculture and Livestock Extension Programme reaches 400,000 farms per year. Many new private enterprises have been formed with the help of government as part of a unique social infrastructure comprising stakeholder forums, implementation teams, focal area development committees and local groups. In this way, public money is being used to build social capital that, in turn, is creating increased productivity of agriculture. The CAR-BAP (African Research Centre on Banana and Plantain) is a good example of a regional research partnership for plantains and bananas across Cameroon, Congo, Côte d'Ivoire, Ghana and Nigeria (Tomekpe and Ganry, 2011). It links researchers, creates novel platforms, undertakes training and disseminates materials. It encourages mass propagation by farmers – after PIF (Plants Issus de Fragments) training, some 10 million new disease-resistant plants were spread to farms in two years.

In Malawi, the fertiliser subsidy programme has been so successful in terms of farmer take-up that net imports of maize fell from 132,000 tonnes at the start of the programme to just 1,000 five years later (Dorward and Chirwa, 2011). The net extra production of maize per hectare has been between 406,000 and 866,000t year. Some 67 per cent of farmers benefited from the receipt of fertiliser coupons (estimated 1.7–2.5 million farmers). The policy is controversial for some, who argue that the money could have been spent differently, that the subsidy is too great a proportion of GDP (now 6.6 per cent) or even that farmers should not have been subsidised (even though OECD countries routinely subsidise their farmers). Yet poor households have seen increases in income of 10–100 per cent (some 60 per cent of maize producers are also buyers of maize, and thus high prices hurt them). In Malawi, there are also 345 fertiliser fallows groups, who have extended practices to some 300,000ha.

In Mali, producer cooperatives for cotton production are now a national priority, and 7,200 have been formed since 2005. In the Oxfam project, organic, conventional and fair-trade cotton cooperatives have led to increased yields, better prices and the adoption of a range of sustainable intensification technologies. Women have particularly benefited from a clear focus on improving their organisation and roles on farms (Traore and Bickersteth, 2011). It is clear that incentives are often needed to help establish and embed novel social and technical infrastructure, so that farmers are able to adopt new practice.

The World Food Programme has used food aid to encourage farmers to adopt conservation agriculture in certain places. In West Africa, aid support has been used to subsidise the initial cultivation of stone bunds and the establishment of nurseries for trees. In other contexts, aid has been used to subsidise FFSs. In the Malawi case above, the nation has spent very large sums on fertiliser subsidies. In every case, there are critics who argue that such external support makes the activity itself inefficient and unlikely to be sustained. The alternative view is that if the subsidy is used to create a new form of social, human or natural capital that will yield benefits over time, or builds capacity in such a way that systems are permanently transformed, then this is an efficient use of public money.

Governments can take further actions to value their own agricultural systems. On average, African countries spend 4–5 per cent of their national budgets on agriculture, compared with 8–14 per cent in Asia (Fan *et al.*, 2009), even though African leaders in 2003 called for a 10 per cent budget allocation to agriculture by 2008–09. In Ghana, when government increased the proportion of the free-on-board (FOB) price of cocoa paid to farmers from 40 to 70 per cent, farmers responded by doubling production, showing what smallholders are capable of achieving when given the appropriate support (Röling, 2010). The proportions of national budgets spent on agriculture still considerably exceed expenditure in industrialised countries on their own agricultural research, development and implementation.

In China, the challenge of agricultural intensification is centred on its unique confluence between demand and resource availability: the Chinese agrifood sector meets most of the food needs of 20 per cent of the world's population, and produces 25 per cent of the world's grain, using less than 9 per cent of the world's arable land, with per capita landholdings amongst the lowest in the world. In 2015, China's grain output reached a historic record of 621.4 Mt following 12 years of relatively continuous growth even though there were serious regional or seasonal droughts in some years. All major crop and livestock products have experienced significant growth over the past three and half decades since the reforms that comm enced in 1978.

This shows that some subsectors in agriculture grew quicker than others. From 1978 to 2015, grain (rice, wheat, maize, beans and tubers) output increased by 104 per cent from 305 Mt to 622 Mt. Meat (pork, beef, lamb) output increased by more than sixfold from 12 Mt in 1980 to 86 Mt in 2015. Vegetables and fruits also increased at a very quick pace. Consequently, the food available to everyone in China was significantly boosted. Per capita availability of grain in China increased from 319 kg in 1978 to 453 kg in 2015, meat from 9 kg to 48 kg, aquatic products from 4.6 kg to 49 kg, while the population increased from 987 million to 1.37 billion in the same period.

Yet, intensification has come at a cost: groundwater pollution from the overuse of nitrate fertilisers, pollution from intensive livestock production, degraded soil, low efficiency of input use and therefore low competiveness. At the same time, food demand is increasing apace, driven by the diet change to a more protein-based diet. Rapid economic growth, urbanisation and market development are challenges, and rising income and urban expansion have boosted the demand for meat, fruit and other non-staple foods.

The major drivers of the agricultural production in the last three and half decades included (i) encouraging policies that mobilised farmers' motivation in agricultural production; (ii) technology progress; (iii) income growth and urbanisation as drivers of both qualitative and quantitative changes in agricultural production;

and (iv) agricultural productivity growth and the increasing use of agricultural inputs (Norse, 2012). However, these driving forces are also bringing about some constraints to the further development of agricultural production. The introduction of household responsibility system was a major driver for agricultural production in the early stage of rural reform. However, small-scale and fragmented household farming plots have become a barrier for further improvement in resource-use efficiency, mechanisation and market competitiveness (Huang and Ding, 2016; Ju *et al.*, 2016). Intensive land and water resource use with high input has caused degradation of the natural resource base and environmental externalities (Norse and Ju, 2015; Lu *et al.*, 2015). In recognition of these challenges, the Chinese government has started implementing a comprehensive strategy to modernise China's agriculture while increasing efficiency and improving environmental outcomes: the zero-growth policy in fertiliser use by 2020 is one of the important components of the strategy, as is land consolidation.

Five key concepts inform China's policy framework for development over the next five years: innovation, coordination, greening, opening up and sharing. The 2016 No. 1 Central Document released on 27 January 2016 (Xinhua, 2016) gives a broad picture of China's strategy for sustainable intensification. This covers the following key aspects:

1　Consolidating the foundation for modern agriculture, enhancing the quality, efficiency and competitiveness of agriculture. China will improve the quality and competitiveness of its agricultural products through high quality farmland and professional farmers catering to the demands of modern agriculture. This will entail developing high quality farmland, advancing irrigation, strengthening innovation and extension systems including for the seed industry, coordinating use of resources and markets at home and abroad and making full use of large family farm operations including by professionalising family farms.

2　Protecting resources and the ecosystem and promoting green agricultural development. This will be achieved by strengthening actions to protect resources and increase efficiency in their utilisation, accelerating the pace at which environmental problems are tackled and, finally, by instituting a food safety strategy.

3　Promoting the integration of primary, secondary and tertiary industries in rural regions and raising farm income by promoting the integrative development of primary, secondary and tertiary industries, logistics, markets and the profit-sharing mechanism in rural regions, including the processing of agricultural products, rural tourism and agro-tourism.

4　Promoting integrated rural–urban development by accelerating the development of rural infrastructure, raising the level of public services, improving rural and agricultural insurance services and encouraging financial institutions to extend credit to agricultural businesses.

Growing the right kinds of support for sustainable intensification

This chapter has shown that there are a number of key requirements for aiding the scaling up of sustainable intensification to large numbers of farmers, in such a way that more people and environments will benefit. It is clear that interventions by government help. Scientific and farmer input into technologies and practices that combine crops–animals with appropriate agroecological and agronomic management. The creation of novel social infrastructure results in both flows of information and builds trust amongst individuals and agencies. And this should result in the improvement of farmer knowledge and capacity through the use of farmer field schools and modern information communication technologies. However, it is also clear that no single project or programme will be able to address all of these at once, and thus the generic need is for an integrated approach that seeks positive synergies over time. Despite great progress, and now the emergence of the term *sustainable intensification* and its component parts, there is still much to be done to ensure agricultural systems worldwide increase productivity fast enough whilst ensuring impacts on natural and social capital are only positive.

Policy support and strong institutional partnerships at every scale are critical. However, in many cases, successes have emerged despite prevailing policies rather than because of them, with exceptions showing how effective policy support can result in significant success and spread. It is also clear that agricultural policy needs to be integrated with other relevant policy areas. Policy support for sustainable intensification of food production will not translate into food security, poverty alleviation or resource conservation if it is not well integrated with appropriate environmental, rural development and economic policies. Without such integration, policies are likely to have short-lived or even perverse impacts.

Now for a final word, on world-building. How far have we got with sustainable intensification; how far still to go? We conclude with a short piece on imagining a future that could be good, and auditing progress to date.

9

WORLD-BUILDING BY REDESIGN

We suggested in the opening chapter that this could be a wonderful world, especially with recent progress towards sustainability in agricultural systems worldwide. There are, however, many problems we have not addressed in detail when considering sustainable intensification. These will continue to make the world less than wonderful. At the consumption end of food systems, well-being worldwide is being challenged by new patterns of over-consumption and old ones of severe hunger. Food suppliers waste much that farmers have struggled and cared over; nutritious foods disposed of for cosmetic or apparent safety reasons. At the farm end of food systems, we must still worry about soil erosion, overuse of pest and disease control compounds, salinisation, loss of genetic diversity in crops and livestock, emissions of polluting gases, use of fossil fuels and pressures on access to water. No one said it was easy.

We conclude this book with some observations on redesign and world-building, and an audit of large-scale changes. We have shown in previous chapters the huge range of innovations being implemented by farmers in countries rich and poor, in systems tropical and temperate, on small farms and large. A thousand flowers are indeed blooming. A key question, though, always centres on scale. One improved farm is cause for celebration: individuals, a family, a community, may be renewed with hope and possibilities for successful futures. But what about all 570 million farmers worldwide; what about all 5 billion hectares of cropland? Some 500 million farmers worldwide each have less than one hectare of land (FAO, 2016c).

One question examined in this book has been simple: is there enough happening worldwide on and around farms to suggest that agriculture can increase productivity whilst improving natural capital? And can this happen at scale, perhaps to affect millions of lives and improve millions of hectares of land? We know there are arguments from some quarters that we do not need to increase agricultural production, that there is enough food if only less were wasted, less energetically inefficient meat consumed and less processed food consumed by the affluent. All would help, but there is no magic wand of redistribution: at particular places, most, if not all, farmers need to produce more and do it by improving environmental services not by damaging or destroying them.

TEEB-AgriFood (2017) call this a theory of change, and note there is not enough information on externalities and the real costs of food, and not enough data on what sustainable intensification can deliver. TEEB also ask: with so much concern about agriculture and food systems, the need to produce more and save natural capital, why not a groundswell for change? They note: "the global agrifood enterprise tolerates little deviances from the commodity-based uniformity of mass produced and processed foods". Paradigmatic change needs vision and hope. It is blocked by fatalism and self-interest, and blind faith in the market. We are locked in to prior choices over technologies, and thus determined by path dependency (Arthur, 1989). Amel *et al.* (2017) also note the need to understand how decision-makers for positive environmental outcomes can help individual transformations at personal level, as well as landscape transformations to improve natural capital and environmental services. We will need to embed learning in policy processes (Young and Esau, 2016), and evolve adaptive governance involved looped learning (Argyris and Schön, 1978).

We have gathered evidence here to show that it is right to be hopeful. Sustainable intensification can work. It has already helped farmers and their families; it has transformed environmental services and natural capital. But will it break through to a mega-scale: will a thousand hectares of innovation become a million; will a thousand farmers engaged in continuous redesign become a million, then ten, a hundred, five hundred million? In China, 100,000 rice farmers have been trained in farmer field schools and have adopted IPM: superb progress. There are still another 120 million rice farmers in China.

When computer game architects construct games for the virtual world, they call their discipline *world-building*. An imaginary world is constructed with coherent history, geography and ecology. People have back-stories and future desires. They are active participants in some kind of story, perhaps ancient or future tales of monsters and thrilling escapes, of dark power, or the clashes between values, the fatal flaws in heroes, the lifting of a great threat: after which life can begin again (McKee, 1998; Booker, 2004). We wonder, then, about our world that needs rebuilding. Do we have the imagination and collective will to overcome climate change, biodiversity loss, pollution of the seas, rampant material consumption alongside persistent hunger, lifestyle-related disease?

The discipline of world-building requires a deep immersive context, an enchantment as Tolkien put it. You know the context in this story: it is not good, yet. You know the geography: it is our planet. Now for redesign.

Agriculture and food are just one economic sector, yet are essential to us all, every day. Now, then, to a new *conculture*, a constructed culture where all our choices around agriculture and food help design a better planet and better lives. In the end, scale matters: four billion hectares of agriculture, seven plus million people. Some world-building is epic in scale, but this one will have to be driven by individual farmers and fishers and their families, facilitators, researchers, policymakers. This is the ultimate redesign challenge set out by Stuart Hill (2014): it is in the end down to you and us.

So: how are we doing with sustainable intensification? There are many terms used to describe approaches that seek both productivity and sustainability outcomes for agriculture. We like sustainable intensification, and we also like diversity in terminology. What matters is not which term is better or more correct than another. What matters is whether each can garner support, and lead to rapid redesign on the ground and in communities. Writing this way about successes at mega-scales could also be dangerous and misguided. Advances are often followed by setbacks: as Andrew Campbell and colleagues (2017) noted about Australian Landcare and land restoration efforts over 40 years: "we are excellent at innovating, we have been equally good at forgetting".

It is also possibly dangerous or just short-sighted to focus on two measures of impact: number of farmers and number of hectares. Most sustainable agriculture and food systems have many impacts: the extraordinary expansion of agroforestry in the Sahel has released women from three hours of daily foraging for wood: now they gather enough in 30 minutes. For them and their families, this will be more important than any sense of a greened or more sustainable environment (Sendzimir *et al.*, 2011). But these two proxies of farmers and hectares to illustrate wider system change will have to do. This is also not a comprehensive audit of SI activities through Efficiency and Substitution routes towards Redesign on all farms. We want to show what is possible, how worlds are indeed being redesigned and rebuilt.

SI has been shown to result in beneficial outcomes for both agricultural output and natural capital. The largest increases in food productivity have occurred in developing countries, mostly starting from a lower output base. In industrialised countries, systems have tended to see increases in efficiency for farmers (lower costs), then reductions in harm to environmental services and improvements to natural capital, and often some reductions in crop and livestock yields. But the global challenge is significant: planetary boundaries are under threat, world population will continue to grow from 7.6 billion (2018) to a stabilisation point of 9–10 billion, and consumption patterns are converging on those typical in affluent countries for some sections of populations, yet still leaving some 800 million people hungry worldwide. A key question centres on scale: is sufficient happening worldwide in farmed landscapes to suggest that agriculture can continue to increase productivity whilst improving natural capital? And can this happen at a scale to affect millions of lives and improve millions of hectares of land, whilst resulting in local reversals of biodiversity loss, reductions in environmental pollution and contamination and reduced greenhouse gas emissions?

We have evaluated SI projects, programmes and initiatives worldwide that have been implemented to scale, using a lower limit of 10^4 farmers or hectares. Forty-seven SI initiatives have exceeded this scale, of which 17 exceed the 10^5 threshold, and 14 the 10^6 scale (Table 9.1; Figures 9.1 and 9.2). We estimate from these projects-initiatives in some 100 countries that 156 M farmers have crossed an important substitution–redesign threshold, and are practising SI on 442 Mha of crop land. This comprises 30.7 per cent of all farmers worldwide; and 8.8 per cent of cropland (total agricultural cropland worldwide = 5 x 10^9 hectares).

TABLE 9.1 Extent of sustainable intensification redesign in 47 initiatives worldwide

Redesign type	Sub-type	Country	Farmer numbers (million)	Hectares under SI (million)
1 Integrated pest management (IPM)	Farmer field schools for integrated pest management	Worldwide, 90 countries in Asia and Africa: especially Indonesia, Philippines, China, Vietnam, Bangladesh, India, Sri Lanka, Nepal, Burkina Faso, Senegal, Kenya	12.00	10.00
	Biological control of pearl millet head miner	Burkina Faso, Niger, Mali, Senegal	0.750	2.00
	Cotton integrated pest management	Egypt	0.150	0.310
	Push–pull IPM	Kenya, Uganda	0.130	0.100
2 Conservation agriculture	Conservation agriculture with zero tillage	Worldwide: Brazil, Argentina, Kazakhstan, USA, Australia, India		
		Industrialised countries	0.450	94.00
		Developing countries	16.50	86.00
	Microbacia groups for watershed management	Brazil, southern: Parana, Santa Catarina	0.100	1.00
	Zai and tassa water harvesting	Burkina Faso, Niger	0.050	0.025
3 Integrated crop and biodiversity redesign	Organic agriculture	Worldwide: especially India, Ethiopia, Mexico (for numbers of farmers)	2.400	51.0
	Rice–fish systems	South East and East Asia	1.00	1.40

(Continued)

TABLE 9.1 (Continued)

Redesign type	Sub-type	Country	Farmer numbers (million)	Hectares under SI (million)
	System of crop intensification for tef	Ethiopia	1.00	1.10
	System of rice intensification	Vietnam	1.00	1.00
	Pigeon pea/maize multiple cropping	East and Southern Africa	0.450	0.250
	Crop redesign with integrated plant and pest management with farmer field schools	Burkina Faso, Niger, Mali, Senegal	0.180	0.150
	Landcare	Australia	0.090	0★
	Campesino a Campesino agroecological farming	Cuba	0.100	0.050
	System of rice intensification	India, all	0.600	0.600
	System of crop intensification	India, Bihar	0.500	0.300
	Zero-budget natural farming	India: Andhra Pradesh	0.125	0.065
	Farmer agroecological wisdom networks	NE Thailand	0.100	0.300
	Science and technology boards	China	0.050	0.030
	Legume–maize intercrops for green manures/ cover crops	Honduras, Guatemala, Mexico, Nicaragua	0.067	0.090
	Green manure/ cover crop mixed systems	Brazil	0.140	0.100
	All crops with mucuna legumes (for *Imperata* suppression)	Benin	0.014	0.030

Redesign type	Sub-type	Country	Farmer numbers (million)	Hectares under SI (million)
	System of crop intensification, millet	India: Madhya Pradesh	0.013	0.013
	Mokichi Okada natural/ nature farming	Japan	0.015	0.003
	Orange-fleshed short-duration sweet potato	Uganda	0.014	0.011
4 Pasture and forage redesign	Management-intensive rotational grazing	USA	0.010	1.60
	Brachiaria-grass mixed crop–forage systems	Brazil	1.30	80.0
	Agropastoral field school	Uganda	0.12	0.25
5 Trees in agricultural systems	Agroforestry and soil conservation	Niger, Burkina Faso, Mali	4.00	3.00
	Joint forest management groups and forest protection committees	India, Nepal	11.6	25.0
	Community-based forestry	Mexico	0.09	15.0
	Forest farmer cooperatives	China, Vietnam	13.80	17.8
	Agroforestry and multifunctional agriculture	Cameroon	0.010	0.005
	Fertiliser and fodder trees and shrubs	Zambia, Malawi	0.500	0.400
6 Irrigation water management	Water user associations for irrigation management	India	15.00	15.00

(*Continued*)

TABLE 9.1 (Continued)

Redesign type	Sub-type	Country	Farmer numbers (million)	Hectares under SI (million)
	Community irrigation management subaks	Indonesia (Bali)	0.900	14.0
	Water users' associations	Mexico	2.00	4.0
7 Intensive small-scale systems	Microcredit group programmes (enablers of small-scale SI): BRAC, Grameen, Proshika	Bangladesh	17.00	8.50
	Intensive vegetable–pig systems with biodigesters	China	50.00	7.00
	Homestead garden production	Bangladesh	0.940	0.010
	Organic small-scale raised beds	Kenya	0.150	0.001
	Allotment gardens	UK	0.300	0.0075
	Community urban gardens	USA and Canada	0.018	0.001
	Group purchasing associations (Community Supported Agriculture, tekei groups, guilds)	USA, France, Japan, Switzerland, Belgium	0.01	0.05
	Integrated aquaculture	Malawi, Cameroon, Ghana	0.018	0.001
Total			**155.7**	**441.5**

Note

We do not present data on adoption of GM crops here, as these have mostly resulted in Efficiency/ Substitution changes: one crop variety for another, some reductions in insecticide, some increases in herbicide, depending on the traits (Frisvold and Reeves, 2014); a number of GM traits are used in conservation agriculture systems.

★ The average farm size in Australia is 3,000 hectares, but there is no data on area under SI within Landcare groups and farms.

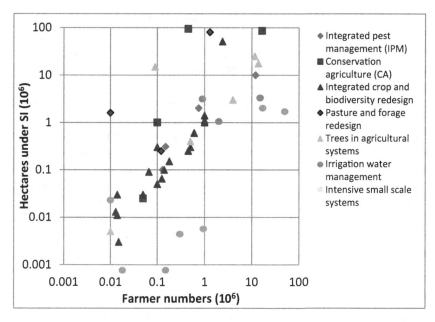

FIGURE 9.1   Farmer numbers and hectares under seven types of sustainable intensification (47 initiatives)

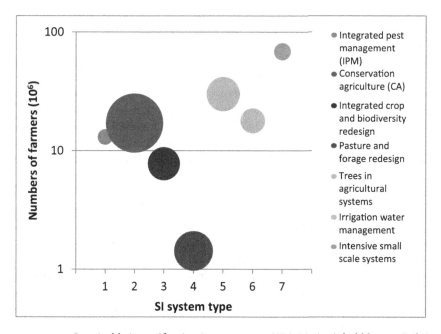

FIGURE 9.2   Sustainable intensification in seven types (47 initiatives): bubble area (Mha)

For SI to have a transformative impact on whole landscapes, it requires cooperation, or at least individual actions that collectively result in wider additive or synergistic benefits. For farmers to be able to adapt their agroecosystems in the face of shocks
and stresses, they will need to have the confidence to innovate. It will be critical
that sustainable intensification does not prescribe concretely defined end points for
technologies and practices. As ecological, climate and economic conditions change,
and as knowledge adapts too, so must the capacity of farmers and communities be
enhanced to allow them to drive transitions, thus implying a process of collective
social learning. Every example of successful redesign for sustainable intensification at
scale has involved the prior building of social capital. There are four features of social
capital: relations of trust; reciprocity and exchange; common rules, norms and sanctions; and connectedness, networks and groups. As social capital lowers the costs of
working together, it facilitates cooperation, and people have the confidence to invest
in collective activities, knowing that others will do so too. They are also less likely to
engage in unfettered private actions that result in resource degradation.

This suggests the need for a new knowledge economy for agriculture in the 21st
century. The technologies and practices increasingly exist to provide both positive
food and ecosystem outcomes: now knowledge needs to be co-created and deployed
in an interconnected fashion. Important examples in industrialised countries include
the Landcare movement in Australia with 6,000 groups, and farmer-led watershed
councils in the USA. These have created platforms for creation of technology and
practice to address locally specific problems, such as erosion, nutrient loss, pathogen
escape and waterlogging. In Cuba, the Campesino a Campesino movement uses
Freirian social communication to integrate agroecology into redesign, with knowledge and technologies spread through exchange, teaching and cooperatives: productivity of the 100,000 farmer members increased by 150 per cent over ten years,
and pesticide use reduced to 15 per cent of former levels. In West Africa, innovation platforms have increased yield and income in maize and cassava systems, and in
Bangladesh have resulted in the development and spread of direct-seeded and early-
maturing rice. In China, Science and Technology Backyard Platforms now operate
in 21 provinces covering a wide range of crops: wheat, maize, rice, soybean, potato,
mango, lychee, vegetable. STBs bring agricultural scientists to live in villages, and use
field demonstrations, farm schools and yield contests to engage farmers in externally
and locally developed innovations. Reasons for success centre more on in-person
communications, socio-cultural bonding and the trust developed amongst farmer
groups of 30–40 individuals. Farmer field schools are already so dense in some locations that they have transformed knowledge co-creation and behavioural change.

The expansion of sustainable intensification has begun to occur at scale across
a wide range of agroecosystems. The benefits of both scientific and farmer input
into technologies and practices that combine crops–animals with appropriate agroecological and agronomic management are increasingly evident. The creation of
novel social infrastructure results in both flows of information and builds trust
amongst individuals and agencies. This should result in the improvement of farmer
knowledge and capacity through the use of platforms for cooperation together

with mobile communication technologies. But state policies for SI remain poorly developed. In the EU, farm subsidies are increasingly targeted towards environmental outcomes, but do not necessarily guarantee the synergistic benefits of whole system SI. A number of countries have offered explicit public policy support to group formation, such as for Landcare (Australia), joint forest management (India, Nepal, DR Congo), irrigation user groups (Mexico) and farmer field schools (Indonesia, Burkina Faso). In Andhra Pradesh, the state government has recently made explicit its support to zero-budget natural farming, aiming to reach 0.5 million farmers in 2,000 villages by 2022. In China, five key concepts are emphasised in the 2016 No. 1 Central Document, with innovation, coordination, greening, opening up and sharing all part of a new strategy for sustainable intensification. At the same time, consumers are increasingly playing a role in connecting directly with farmers, such as through group purchasing schemes, farmers' markets, on-farm stores and certification schemes, which may in turn change consumption behaviours.

Sustainable intensification represents a simple set of principles and a broad range of approaches and methods, each of which should emerge from place, culture and people. With farmers leading the way, supported by other actors in the agricultural and food knowledge economy, who knows, perhaps the whole world could be rebuilt?

For there to be substantial and noticeable impact at planetary level, we will need 5,000 projects of 1 Mha; we will need 7,000 activities that touch more than 1 million people at a time. World-building is daunting. On the other hand, adoption of a SI approach on 100 Mha would mean only another 49 such innovations would be required (to cover the world's 5 billion hectares of agriculture). There are 570 million farmers worldwide: a batch of 100 redesigns at 5 M farmers apiece covers them all.

There is much to be done to ensure agricultural systems worldwide increase productivity fast enough whilst ensuring impacts on natural and social capital are positive. We conclude a transition from efficiency through substitution to redesign will be essential, suggesting that the concept and practice of the sustainable intensification of agriculture will be a journey of adaptation and improvement, driven by a wide range of actors cooperating in a new agricultural knowledge economy.

One review of the shift to management-intensive rotational grazing in Wisconsin showed that farmers could earn more, there were time savings, environmental benefits of reduced erosion and nutrient-loss, on-farm wildlife benefits, improved animal health and welfare, and increased productivity: above all, farmers said this was "a peaceful way of farming" (Undersander et al., 2002). Stuart Hill (2015) noted that wise farmers tended to be calmer, more attentive and interested in what was going on around them, especially in nature.

These are the greater goals in life: well-being, contentment, meaning. Confucius did write: "If your plan is for one year, plant rice; if your plan is for ten years, plants trees; if your plan is for one hundred years, educate children." This is beginning to look like the emergence of an effective knowledge economy built around sustainable intensification for the 21st century and beyond.

REFERENCES

Adhikari, P., Araya, H., Aruna, G., Balamatti, A., Banerjee, S., Baskaran, P., *et al.* 2018. System of crop intensification for more productive, resource-conserving, climate-resilient, and sustainable agriculture: experience with diverse crops in varying agroecologies. *International Journal of Agricultural Sustainability*, pp. 1–28.

Africare, Oxfam America and WWF–ICRISAT Project. 2010. *More Rice for People, More Water for the Planet.* WWF–ICRISAT Project, Hyderabad, India.

Ajayi, C., Akinnifesi, F., Sileshi, G. and Kanjipite, W. 2009. Labour inputs and financial profitability of conventional and agroforestry-based soil fertility management practices in Zambia. *Agrekon* 48: 246–292.

Ali, K. 2017. Hydroponic Farming Takes Root in CT. 1 May. https://www.igrow.news/home-1/hydroponic-farming-takes-root-in-ct

Ali, M.Y., Waddington, S.R., Hodson, D., Timsina, J. and Dixon, J. 2009. *Maize–rice cropping systems in Bangladesh: status and research opportunities.* Working Paper, Mexico DF: CIMMYT.

Alinovi, L., Hemrich, G. and Russo, L. 2008. *Beyond Relief: Food Security in Protracted Crises.* Rugby, UK: Practical Action Publishing.

Altieri, M. 1995. *Agroecology: The Science of Sustainable Agriculture.* Boulder, CO: Westview.

Altieri, M.A. and Toledo, V.M. 2011. The agroecological revolution in Latin America: rescuing nature, ensuring food sovereignty and empowering peasants. *Journal of Peasant Studies* 38: 587–612.

AMAP (Associations pour le maintien d'une agriculture paysanne). 2017. www.reseau-amap.org

Amel, E., Manning, C., Scott, B. and Koger, S. 2017. Beyond the roots of human inaction: fostering collective effort toward ecosystem conservation. *Science* 356: 275–279.

American Community Gardening Association. 2014. https://communitygarden.org/resources/faq/

Anas, I., Rupela, O.P., Thiyagarajan, T.M. and Uphoff, N. 2011. A review of studies on SRI effects on beneficial organisms in rice soil rhizospheres. *Paddy Water Environment* 9: 53–64.

Andersson, H., Tago, D. and Treich, N. 2014. *Pesticides and health: a review of evidence on health effects, valuation of risks, and benefit-cost analysis.* Toulouse School of Economics, Working Paper TSE-477. Toulouse Cedex 6, France.

Angus, J.F., Kirkegaard, J.A., Hunt, J.R., Ryan, M.H., Ohlander, L. and Peoples, M.B. 2015. Break crops and rotations for wheat. *Crop and Pasture Science* 66(6): 523–552.

Anthony, D. 2005. Cooperation in microcredit borrowing groups: identity, sanctions, and reciprocity in the production of collective goods. *American Sociological Review* 70(3): 496–515.

Argyris, C. and Schön, D.A. 1978. *Organizational Learning: A Theory of Action Perspective*. Reading, MA: Addison-Wesley.

Armitage, D., Marschke, M. and Plummer, R. 2008. Adaptive co-management and the paradox of learning. *Global Environmental Change* 18(1): 86–98.

Arthur, W.B. 1989. Competing technologies, increasing returns, and lock-in by historical events. *Economic Journal* 99(394): 116–131.

Assefa, K., Aliye, S., Belay, G., Metaferia, G., Tefera, H. and Sorrells, M.E. 2011. *Quncho*: the first popular tef variety in Ethiopia. *International Journal of Agricultural Sustainability* 9(1): 25–34.

Assefa, T., Sperling, L., Dagne, B., Argaw, W., Tessema, D. and Beebe, S. 2014. Participatory plant breeding with traders and farmers for white pea bean in Ethiopia. *Journal of Agricultural Education and Extension* 20(5): 497–512.

Atangana, A., Khasa, D., Chang, S. and Degrande, A. 2014. Major agroforestry systems of the humid tropics. In: Atangana, A., Khasa, D., Chang, S. and Degrande, A. (eds) *Tropical Agroforestry*. Dordrecht: Springer.

Athukorala, W., Wilson, C. and Robinson, T. 2010. Determinants of health costs due to farmers' exposure to pesticides: an empirical analysis. In: *Proceedings of the 85th Annual Conference of Western Economic Association*, Portland, Oregon, 29 June–3 July.

Ayarza, M.A. and Welchez, L.A. 2004. Drivers effecting the development and sustainability of the Quesungual Slash and Mulch Agroforestry System (QSMAS) on hillsides of Honduras. Comprehensive Assessment Bright Spots Project. Final Report. URL: https://core.ac.uk/download/pdf/41712381.pdf

Baker, J.M., Ochsner, T.E., Venterea, R.T. and Girffis, T.J. 2007. Tillage and soil carbon sequestration – What do we really know? *Agriculture, Ecosystems & Environment* 118: 1–5.

Barański, M., Średnicka-Tober, D., Volakakis, N., Seal, C., Sanderson, R., Stewart G.B., *et al.* 2014. Higher antioxidant and lower cadmium concentrations and lower incidence of pesticide residues in organically grown crops: a systematic literature review and meta-analyses. *British Journal of Nutrition* 112(5): 794–811.

Barbier, E.B. 2015. Redressing the structural imbalance. In: Barbier, E.B. *Nature and Wealth* (pp. 165–183). Basingstoke: Palgrave Macmillan.

Barison, J. and Uphoff, N. 2011. Rice yield and its relation to root growth and nutrient-use efficiency under SRI and conventional cultivation: an evaluation in Madagascar. *Paddy and Water Environment* 9: 65–78.

Barnard, A. 1999. Images of hunters and gatherers in European social thought. In: Lee, R.B. and Daly, R. (eds) *Cambridge Encyclopedia of Hunters and Gatherers* (pp. 375–383). Cambridge: Cambridge University Press.

Barzman, M., Bertschinger, L., Dachbrodt-Saaydeh, S., Graf, B., Jensen, J.E., Joergensen, L.N., *et al.* 2014. *Integrated Pest Management: Experiences with Implementation, Global Overview*. Dordrecht: Springer.

Bawden, R.J. 1998. The community challenge: the learning response. *New Horizons in Education* 99: 40–59.

Baynes, J., Herbohn, J., Smith, C., Fisher, R. and Bray, D. 2015. Key factors which influence the success of community forestry in developing countries. *Global Environmental Change* 35: 226–238.

Bebber, D.P., Ramotowski, M.A.T. and Gurr, S.J. 2013. Crop pests and pathogens move polewards in a warming world. *Nature Climate Change* 3: 985–988.

Behera, B. and Engel, S. 2006. Institutonal analysis of evolution of joint forest management in India: a new institutional economics approach. *Forest Policy and Economics* 8: 350–362.

Benbrook, C. 2012. Impacts of genetically engineered crops on pesticide use in the U.S. – the first sixteen years. *Environmental Sciences Europe* 24: 24 doi: 10.1186/2190-4715-24-24

Bennett, E.M., Solan, M., Biggs, R., McPhearson, T., Norström, A.V., Olsson P., *et al.* 2016. Bright spots: seeds of a good Anthropocene. *Frontiers in Ecology and the Environment* 14(8): 441–448.

Bennett, M. and Franzel, S. 2013. Can organic and resource-conserving agriculture improve livelihoods? A synthesis. *International Journal of Agricultural Sustainability* 11(3): 193–215.

Bentley, J.W. 2009. Impact of IPM extension for smallholder farmers in the tropics. In: Peshin, R. and Dhawan, A.K. (eds) *Integrated Pest Management: Dissemination and Impact.* Berlin: Springer.

Berdegué, J.E. and Escobar, G. 2001. Agricultural knowledge and information systems and poverty reduction. *World Bank Discussion Paper.* URL: http://rimisp.org/wp-content/uploads/2013/06/0115-000824-akisandpovertyrevisedfinal.pdf

Berkes, F. 2017. *Sacred Ecology.* Fourth Edition. London: Routledge.

Berkes, F. and Ross, H. 2013. Community resilience: toward an integrated approach. *Society & Natural Resources* 26(1): 5–20.

Berry, W. 2010. *What Matters? Economics for a Renewed Commonwealth.* Berkeley, CA: Counterpoint.

Bharucha, Z.P. 2014. The Brazil–China Soy Complex: A Global Link in the Food–Energy–Climate Change Trilemma. Centre for Research in Economic Sociology and Innovation (CRESI) Working Paper 2014–01, University of Essex, Colchester, UK. URL: http://repository.essex.ac.uk/10510/

Bharucha, Z.P. and Pretty, J.N. 2010. The role and importance of wild foods in agricultural systems. *Philosophical Transactions of the Royal Society B* 365: 2913–2926.

Bharucha, Z.P., Smith, D. and Pretty, J. 2014. All paths lead to rain: explaining why watershed development in India does not alleviate the experience of water scarcity. *Journal of Development Studies* 50(9) doi: 10.1080/00220388.2014.928699

Bianchi, F.J., Booij, C.J.H. and Tscharntke, T. 2006. Sustainable pest regulation in agricultural landscapes: a review on landscape composition, biodiversity and natural pest control. *Proceedings of the Royal Society of London B: Biological Sciences* 273(1595): 1715–1727.

Bioversity International. 2016. Mainstreaming Agrobiodiversity in Sustainable Food Systems: Scientific Foundations for an Agrobiodiversity Index – Summary. Bioversity International, Rome, Italy.

Birch, A.N.E, Begg, G.S. and Squire, G.R. 2011. How agro-ecological research helps to address food security issues under new IPM and pesticide reduction policies for global crop production systems. *Journal of Experimental Biology* 62(10): 3251–3261.

Blomley, T. 2013. *Lessons learned from community forestry in Africa and their relevance for REDD+.* Washington, DC, Forest Carbon, Markets and Communities Program.

Blomquist, W. and Schlager, E. 2005. Political pitfalls of integrated watershed management. *Society and Natural Resources* 18(2): 101–117.

Boehm, S., Bharucha, Z.P. and Pretty, J.N. (eds) 2014. *Ecocultures: Blueprints for Sustainable Communities.* London: Routledge.

Bohra, A., Pandey, M.K., Jha, U.C., Singh, B., Singh, I.P., Datta, D., *et al.* 2014. Genomics-assisted breeding in four major pulse crops of developing countries: present status and prospects. *Theoretical and Applied Genetics* 127(6): 1263–1291.

Booker, C. 2004. *The Seven Basic Plots: Why We Tell Stories.* London: Bloomsbury.

Bottrell, D.G. and Schoenly, K.G. 2012. Resurrecting the ghost of green revolutions past: the brown planthopper as a recurring threat to high-yielding rice production in tropical Asia. *Journal of Asia-Pacific Entomology* 15: 122–140.

Bouis, H. 1996. Enrichment of food staples through plant breeding: a new strategy for fighting micronutrient malnutrition. *Nutrition Reviews* 54(5): 131–137.

Braun, A. and Duveskog, D. 2009. *The Farmer Field School Approach – History, Global Assessment and Success Stories*. Rome: IFAD.

Brenna, S., Rocca, A. and Sciaccaluga, M. 2013. Stock di carbonio organico e fertilità biologica. In: Il ruolo dell'Agricoltura Conservativa nel bilancio del carbonio – progetto Agricoltura. Regione Lombardia, QdR n. 153/2013, cap. 3.1: 53–74.

Brewer, M.J. and Goodell, P.B. 2012. Approaches and incentives to implement integrated pest management that addresses regional and environmental issues. *Annual Review of Entomology* 57: 41–59.

Brouder, S.M. and Gomez-MacPherson, H. 2014. The impact of conservation agriculture on smallholder agricultural yields: a scoping review of the evidence. *Agriculture, Ecosystems and Environment* 187: 11–32.

Brummett, R.E. and Jamu, D.M. 2011. From researcher to farmer: partnerships in integrated aquaculture–agriculture systems in Malawi and Cameroon. *International Journal of Agricultural Sustainability* 9(1): 282–289.

Buckwell, A., Nordang Uhre, A., Williams, A., Polakova, J., Blum, W.E.H., Schiefer, J., *et al.* 2014. *The Sustainable Intensification of European Agriculture*. Brussels: RISE Foundation.

Bulte, E., Boone, R.B., Stringer, R. and Thornton, P.K. 2008. Elephants or onions? Paying for nature in Amboseli, Kenya. *Environment and Development Economics* 13(3): 395–414.

Bunch, R, 2018 (2012). *Restoring the Soil*. Canadian Foodgrains Bank, Winnipeg.

Calvet-Mir, L., Gómez-Baggethun, E. and Reyes-García, V. 2012. Beyond food production: ecosystem services provided by home gardens. A case study in Vall Fosca, Catalan Pyrenees, Northeastern Spain. *Ecological Economics* 74: 153–160.

Cambardella, C.A., Delate, K. and Jaynes, D.B. 2015. Water quality in organic systems. *Sustainable Agriculture Research* 4(3): 60.

Campbell, A. 1994. *Landcare: Communities Shaping the Land and the Future*. Sydney: Allen and Unwin.

Campbell, A., Alexandra, J. and Curtis, D. 2017. Reflections on four decades of land restoration in Australia. *Rangeland Journal* 39(6): 405–416. https://doi.org/10.1071/RJ17056

Campbell, B., Beare, D., Bennett, E., Hall-Spencer, J., Ingram, J., Jaramillo, F., *et al.* 2017. Agriculture production as a major driver of the Earth system exceeding planetary boundaries. *Ecology and Society* 22(4).

Carolan, M.S. 2013. *Reclaiming Food Security*. London: Routledge.

Carson, R. 1962. *Silent Spring*. Boston, MA: Houghton Mifflin.

Cary, J. and Web, T. 2000. *Community Landcare, the National Landcare Program and the Landcare Movement: The Social Dimensions of Landcare*. Kingston, ACT: Bureau of Rural Science.

Cato, M.P. 1979. Di agri cultura. In: Hooper, W.D. (revised Ash, H.B.). *Marcus Porcius Cato on Agriculture*. Cambridge, MA: Harvard University Press.

Chamberlain, D.E., Fuller, R.J., Bunce, R.G.H., Duckworth, J.C. and Shrubb, M. 2000. Changes in the abundance of farmland birds in relation to the timing of agricultural intensification in England and Wales. *Journal of Applied Ecology* 37(5): 771–788.

Chambers, R. 1983. *Rural Development: Putting the Last First*. Harlow: Prentice Hall.

Chantre, E. and Cardona, A. 2014. Trajectories of French field crop farmers moving toward sustainable farming practices: change, learning, and links with the advisory services. *Agroecology and Sustainable Food Systems* 38(5): 573–602.

Charles, R. and Vuilloud, P. 2001. Pois protéagineux et azote dans la rotation. *Revue Suisse d'Agriculture* 33: 365–370.

Chhatre, A. and Agrawal, A. 2009. Trade-offs and synergies between carbon storage and livelihood benefits from forest commons. *Proceedings of the National Academy of Sciences* 106(42): 17667–17670.

Chhay, N., Seng, S., Tanaka, T., Yamauchi, A., Cedicol, E.C., Kawakita, K. and Chiba, S. 2017. Rice productivity improvement in Cambodia through the application of technical recommendation in a farmer field school. *International Journal of Agricultural Sustainability* 15(1): 54–69.

Chiffoleau, Y. and Desclaux, D. 2006. Participatory plant breeding: the best way to breed for sustainable agriculture? *International Journal of Agricultural Sustainability* 4(2): 119–130.

China Rural Statistical Report. 2013. China Statistic Press, Beijing, China.

Christanty, L., Abdoellah, O.S., Marten, G.G. and Iskander, J. 1986. Traditional agroforestry in West Java: the pekarangan (homegarden) and kebun-talun (annual-perennial rotation) cropping systems. In: Marten, G.G. (ed.) *Traditional Agriculture in Southeast Asia: A Human Ecology Perspective.* Boulder, CO: Westview Press.

CIAT (International Centre for Tropical Agriculture). 2009. Quesungual slash and mulch agroforestry system (QSMAS): improving crop water productivity, food security and resource quality in the sub-humid tropics. Project Number 15. Cali, Colombia: CIAT. URL: https://cgspace.cgiar.org/bitstream/handle/10568/3906/PN15_CIAT_Project%20Report_Jun09_final.pdf?sequence=1

CIAT. 2013. *The Impacts of CIAT's Collaborative Research.* Cali, Colombia: CIAT.

CIMMYT. 2013. Water-saving techniques salvage wheat in drought-stricken Kazakhstan. In: *Wheat Research, Asia.* 21 March. www.cimmyt.org/water-saving-techniques-salvage-wheat-in-drought-stricken-kazakhstan

CMO (Chief Medical Officer). 2013. *Chief Medical Officer's Annual Report 2012: Our Children Deserve Better: Prevention Pays.* London: UK Government.

Coleman, J. 1988. Social capital and the creation of human capital. *American Journal of Sociology* 94: S95–S120.

Collier, W.L., Wiradi, G. and Soentoros 1973. Recent changes in rice harvesting methods. Some serious social implications. *Bulletin of Indonesian Economic Studies* 9(2): 36–45.

Conway, G.R. 1997. *The Doubly Green Revolution.* London: Penguin.

Conway, G.R. 2012. *One Billion Hungry: Can We Feed the World?* Ithaca, NY: Comstock Publishing.

Conway, G.R. and Barbier, E.B. 1990. *After the Green Revolution. Sustainable Agriculture for Development.* London: Earthscan.

Conway, G.R. and Pretty, J. 1991. *Unwelcome Harvest: Agriculture and Pollution.* London: Earthscan. Cook, S.M., Khan, Z.R. and Picket, J.A. 2007. The use of push–pull strategies in integrated pest management. *Annual Review of Entomology* 52: 375–400.

Costanza, R., d'Arge, R., de Groot, R., Farber, S., Grasso, M., Hannon, B., *et al.* 1997. The value of the world's ecosystem services and natural capital. *Nature* 387: 253–260.

Costanza, R., de Groot, R., Sutton, P., van der Ploeg, S., Anderson, S.J., Kubiszewski, I., *et al.* 2014. Changes in the global value of ecosystem services. *Global Environmental Change* 26: 152–158.

Council of the European Union. 2017. Green light to new European rules on organic farming. Press Release 421/17. 28/06/2017 URL: www.consilium.europa.eu/en/press/press-releases/2017/06/28/rules-organic-farming/pdf

Crews, T.E. and Brookes, P.C. 2014. Changes in soil phosphorus forms through time in perennial versus annual agroecosystems. *Agriculture, Ecosystems and Environment* 184: 168–181.

Crouch, D. and Ward, C. 1997. *The Allotment: Its Landscape and Culture.* Nottingham: Five Leaves Publications.

Curtis, A., Ross, H., Marshall, G.R., Baldwin, C., Cavaye, J., Freeman, C., *et al.* 2014. The great experiment with devolved NRM governance: lessons from community engagement in Australia and New Zealand since the 1980s. *Australasian Journal of Environmental Management* 21(2): 175–199.

Daly, H. and Cobb, J.B. 1989. *For the Common Good: Redirecting the Economy towards Community, the Environment and Sustainable Development.* London: Green Print.

DAMM. 2012. *Study on the technology of energy-saving and emission-reducing of agricultural mechanization.* Beijing: China Agricultural Science and Technology Press.

Dangour, A.D., Dodhia, S.K., Hayter, A., Allen, E., Lock, K. and Uauy, R. 2009. Nutritional quality of organic foods: a systematic review. *American Journal of Clinical Nutrition* 90: 680–685.

Dasgupta, P. 2010. Nature's role in sustaining economic development. *Philosophical Transactions of the Royal Society of London B: Biological Sciences* 365(1537): 5–11.

Davy, B.J., Kish, K., Zywert, K. and Quilley, S. 2017. Environmentalism at the margins: exploring existing possibilities for an alternative modernity. Paper presented at Resilience 2017. Stockholm, 20–23 August.

De Bon, H., Huat, J., Parrot, L., Sinzogan, A., Martin, T., Malézieux, E. and Vayssieres, J. 2014. Pesticide risks from fruit and vegetable pest management by small farmers in sub-Saharan Africa. A review. *Agronomy for Sustainable Development* doi: 10.1007/s13593-014-0216-7

De Schutter, O. and Vanloqueren, G. 2011. The new green revolution: how twenty-first-century science can feed the world. *Solutions* 2(4): 33–44.

Defra. 2012. Green Food Project Conclusions. Defra. URL: https://www.gov.uk/government/uploads/system/uploads/attachment_data/file/69575/pb13794-greenfoodproject-report.pdf

Defra. 2014. Pesticide usage statistics. URL: https://secure.fera.defra.gov.uk/pusstats/

Derpsch, R. 2008. No-tillage and conservation agriculture: a progress report. In: Goddard, T., Zoebisch, M.A., Gan, Y.T., Ellis, W., Watson, A. and Sombatpanit, S. (eds) *No-Till Farming Systems.* Special Publication No. 3. Bangkok: World Association of Soil and Water Conservation.

Derpsch, R. 2014. Why do we need to standardize no-tillage research? *Soil and Tillage Research* 137: 16–22.

Derpsch, R., Friedrich, T., Kassam, A. and Hongwen, L. 2010. Current status and adoption of no-till farming in the world and some of its main benefits. *International Journal of Agriculture and Biological Engineering* 3: 1–25.

Derpsch, R., Lange, D., Birbaumer, G. and Moriya, K. 2016. Why do medium- and large-scale farmers succeed practicing CA and small-scale farmers often do not? Experiences from Paraguay. *International Journal of Agricultural Sustainability* 14(3): 269–281.

Detroit Food and Fitness Collaborative. 2014. URL: http://detroitfoodandfitness.com/

Dewar, J.A. 2007. Perennial polyculture farming: seeds of another agricultural revolution? Santa Monica, CA: RAND Corporation.

Dhar, S., Barah, B.C., Vyas, A.K. and Uphoff, N. 2016. Comparing System of Wheat Intensification (SWI) with standard recommended practices in the north-western plain zone of India. *Archives of Agronomy and Soil Science* 62: 994–1006.

Dietz, R. and O'Neill, D.W. 2013. *Enough is Enough. Building a Sustainable Economy in a World of Finite Resources.* London: Routledge.

Dietz, T. and Rosa, E.A. 1994. Rethinking the environmental impacts of population, affluence and technology. *Human Ecology Review* 1: 277–300.

Dixon, J. and Gulliver, J., with Gibbon, D. 2001. Farming Systems and Poverty. Improving farmer's livelihoods in a changing world. FAO and the World Bank. Rome and Washington, DC. www.fao.org/3/a-ac349e.pdf

Dobbs, T.L. and Pretty, J.N. 2004. Agri-environmental stewardship schemes and 'multifunctionality'. *Review of Agricultural Economics* 26(2): 220–237.

Dobermann, A.A. 2004. A critical assessment of the system of rice intensification (SRI). *Agricultural Systems* 79(3): 261–281.

Dong, W.X., Hu, C.S., Chen, S.Y. and Zhang, Y.M. 2009. Tillage and residue management effects on soil carbon and CO2 emission in a wheat-corn double-cropping system. *Nutrient Cycling in Agroecosystems* 83: 27–37.

Dorward, A. and Chirwa, E. 2011. The Malawi agricultural input subsidy programme: 2005/06 to 2008/09. *International Journal of Agricultural Sustainability* 9(2): 232–247.

Dudley, N., Attwood, S., Goulson, D., Jarvis, D., Bhuarucha, Z. and Pretty, J. 2017. How should conservationists respond to pesticides as a driver of biodiversity loss in agroecosystems? *Biological Conservation* 209: 449–453.

Dũng, N.T. *et al.* 2011. Simple and effective: SRI and agricultural innovation. URL: http://sri.cals.cornell.edu/countries/vietnam/VN_SRI_booklet_Eng2012.pdf

Ehler, L.E. 2006. Integrated pest management (IPM): definition, historical development and implementation, and the other IPM. *Pest Management Science* 62: 787–789.

Ehrlich, P. and Ehrlich, A. 1968. *The Population Bomb.* New York: Sierra Club/Ballantine Books.

Ehrlich, P.R. and Ehrlich, A. H. 2013. Can a collapse of global civilization be avoided? *Proceedings of the National Academy of Sciences B* 280: 20122854.

Eilers, E., Kremen, C., Greenleaf, S.S., Garber, A.K. and Klein, A. 2011. Contribution of pollinator-mediated crops to nutrients in the human food supply. *PLOS One* 6(6). doi: 10.1371/journal.pone.0021363

Elliot, J., Firbank, L.G., Drake, B., Cao, Y. and Gooday, R. 2013. Exploring the concept of sustainable intensification. URL: www.snh.gov.uk/docs/A931058.pdf

Ellis, E.C. and Wang, S.M. 1997. Sustainable traditional agriculture in the Tai Lake Region of China. *Agriculture, Ecosystems & Environment* 61(2–3): 177–193.

EPA. 2007. Pesticide Industry Sales and Usage: 2006 and 2007 Market Estimates. Washington, DC: Environment Protection Agency.

Eswaran, H., Van den Berg, E., Reich, P. and Kimble, J.M. 1995. Global soil C resources. In: Lal, R., Kimble, J., Levine, E. and Stewart, B.A. (eds) *Soils and Global Change* (pp. 27–43). Boca Raton, FL: Lewis Publishers.

European Bird Census Council. 2015. European Wild Bird Indicators, 2015 update. URL: www.ebcc.info/index.php?ID=588

Fan, S., Omilola, B. and Lambert, M. 2009. Public Spending for Agriculture in Africa: Trends and Composition. ReSAKSS Working Paper 28. URL: http://citeseerx.ist.psu.edu/viewdoc/download?doi=10.1.1.210.3668&rep=rep1&type=pdf

FAO. 2000. Conflicts, agriculture and food security. URL: www.fao.org/docrep/x4400e/x4400e07.htm

FAO. 2004. *Culture of Fish in Rice Fields.* M. Halwart and M. Gupta (eds). Rome: FAO.

FAO. 2007. Analysis of feeds and fertilizers for sustainable aquaculture development in China. Miao, W.M. and Mengqing, L. In: Hasan, M., Hecht, T. and De Silva, S. (eds) *Study and Analysis of Feeds and Fertilizers for Sustainable Aquaculture Development.* FAO Fisheries Technical Paper 497. Rome: FAO.

FAO. 2011. *Save and Grow: A Policymaker's Guide to the Sustainable Intensification of Smallholder Crop Production.* Rome: FAO.

FAO. 2013a. Food wastage footprint: Full-cost accounting. Final report. URL: www.fao.org/3/a-i3991e.pdf

FAO. 2013b. *Climate-Smart Agriculture Sourcebook.* Rome: FAO.

FAO. 2013c. *Supporting Communities in Building Resilience through Agro-pastoral Field Schools.* Rome: FAO.

FAO. 2013d. Save and grow: Cassava: policy brief. Rome: FAO.

FAO. 2013e. *The State of Food and Agriculture. Food Systems for Better Nutrition.* Rome: FAO.

FAO. 2014a. FAOStat. Online statistical database. FAO.

FAO. 2014b. *Aquatic biodiversity in rice-based ecosystems: Studies and reports from Indonesia, LAO PDR and the Philippines.* M. Halwart and D. Bartley (eds). The Asia Regional Rice Initiative: Aquaculture and fisheries in rice-based ecosystems. Rome: FAO.

FAO. 2016a. Report from the FAO regional policy dialogue on ecosystem services from sustainable agriculture for biodiversity conservation. Nairobi, Kenya, 25–26 May. URL: www.fao.org/3/a-bp441e.pdf

FAO. 2016b. *State of the World's Forests. Forests and Agriculture: Land-Use Challenges and Opportunities.* Rome: FAO.

FAO. 2016c. *Save and Grow: Maize, Rice and Wheat – A Guide to Sustainable Crop Production.* Rome: FAO.

FAO. 2016d. *Farmer Field School Guidance Document.* Rome: FAO.

FAO. 2016e. *The State of World Fisheries and Aquaculture.* Rome: FAO.

FAO. 2016f. *Forty Years of Community-Based Forestry.* Rome: FAO.

FAO. 2017a. *Soil Organic Carbon.* Rome: FAO.

FAO. 2017b. *State of Food Security and Nutrition in the World.* Rome: FAO.

FAO. 2017c. *The Future of Food and Agriculture. Trends and Challenges.* Rome: FAO.

FAOSTAT. 2017. Food and Agriculture Organisation (FAO) statistics database. Rome: FAO.

Farooq, M., Siddique, K.H.M., Rehman, H., Aziz, T., Lee, D. and Wahid, A. 2011. Rice direct seeding: experiences, challenges and opportunities. *Soil and Tillage Research* 111: 87–98.

Feder, G., Murgai, R. and Quizon, J.B. 2004. Sending farmers back to school: the impact of Farmer Field Schools in Indonesia. *Review of Agricultural Economics* 26(1): 45–62.

Fedoroff, N.V. 2010. Radically rethinking agriculture for the 21st century. *Science* 327: 833–834.

Field, R.H., Hill, R.K., Carroll, M.J. and Morris, A.J. 2016. Making explicit agricultural ecosystem service tradeoffs: a case study of an English lowland arable farm. *International Journal of Agricultural Sustainability* 14(3): 249–268.

Firbank, L., Bradbury, R.B., McCracken, D.I. and Stoate, C. 2011. Delivering multiple ecosystem services from Enclosed Farmland in the UK. *Agriculture, Ecosystems and Environment* 166: 65–75.

Fisher, L. and Newig, J. 2016. Importance of actors and agency in sustainability transitions: a systematic exploration of the literature. *Sustainability* 8(5): 476. doi: 10.3390/su8050476

FLWC. 2015. Getting started with farmer-led Watershed Councils. URL: https://blogs.ces.uwex.edu/wflcp/files/2015/12/FLWC-GettingStarted.pdf

Foresight. 2011. *The Future of Food and Farming: Challenges and Choices for Global Sustainability.* Final Project Report. London: Government Office for Science.

Franzluebbers, A.J. 2010. Achieving soil organic carbon sequestration with conservation agricultural systems in the southeastern United States. *Soil Science Society of America Journal* 74: 347–357.

Frayne, B., Crush, J. and McLachlan, M. 2014. Urbanization, nutrition and development in Southern African cities. *Food Security* 6(1): 101–112.

Freire, P. 1970. *Pedagogy of the Oppressed.* New York: Continuum.

Friis-Hansen, E. 2012. The empowerment route to well-being: an analysis of farmer field schools in East Africa. *World Development* 40(2): 414–427.

Frisvold, G.B. and Reeves, J.M. 2014. Resistance management and sustainable use of agricultural biotechnology. *AgBioForum* 13(4): 343–359.

Fuhlendorf, S.D. and Smeins, F.E. 1999. Scaling effects of grazing in a semi-arid savanna. *Journal of Vegetation Science* 10: 731–738.

Fukuoka, M. 1985. *The One-Straw Revolution: An Introduction to Natural Farming.* Trans. Pearce, C., Kurosawa, T. and Korn, L. Emmaus, PA: Rodale Press.

Gadanakis, Y., Bennett, R., Park, J. and Areal, F.J. 2015. Evaluating the sustainable intensification of arable farms. Journal of Environmental Management 150: 288–298.

Galhena, D.H., Freed, R. and Maredia, K.M. 2013. Home gardens: a promising approach to enhance food security and wellbeing. *Agriculture & Food Security* 2: 8.

Galluzzi, G., Eyzaguirre, P. and Negri, V. 2010. Home gardens: neglected hotspots of agrobiodiversity and cultural diversity. *Biodiversity and Conservation* 19: 3635–3654.

Gandhi, R., Veeraraghavan, R., Toyama, K. and Ramprasad, V. 2009. Digital green: participatory video for agricultural extension. *Information Technologies and International Development* 5(1): 1–15.

Garbach, K., Milder, J.C., DeClerck, F.A., Montenegro de Wit, M., Driscoll, L. and Gemmill-Herren, B. 2017. Examining multi-functionality for crop yield and ecosystem services in five systems of agroecological intensification. *International Journal of Agricultural Sustainability* 15(1): 11–28.

Garnett, T. 2000. Urban agriculture in London: rethinking our food economy. In: Bakker, N., Dubbeling, M., Gündel, S., Sabel-Koschella, U. and de Zeeuw, H. (eds) *Growing Cities, Growing Food: Urban Agriculture on the Policy Agenda. A Reader on Urban Agriculture*. Feldafing, Germany: German Foundation for International Development.

Garnett, T. and Godfray, C.H. 2012. Sustainable intensification in agriculture: navigating a course through competing food system priorities. Food Climate Research Network and the Oxford Martin Programme on the Future of Food, University of Oxford, UK.

Garrity, D.P., Akinnifesi, F.K., Ajayi, O.C., Weldesemayat, S.G., Mowo, J.G., Kalinganire, A., *et al.* 2010. Evergreen agriculture: a robust approach to sustainable food security in Africa. *Food Security* 2: 197–214.

Gathala, M.K., Timsina, J., Islam, Md. S., Rahman, Md. M., Hossain, Md. I., Harun-Ar-Rashid, Md., *et al.* 2014. Conservation agriculture based tillage and crop establishment options can maintain farmers' yields and increase profits in South Asia's rice–maize systems: evidence from Bangladesh. *Field Crops Research* 172: 85–98.

Giller, K.E., Witter, E., Corbeels, M. and Tittonell, P. 2009. Conservation agriculture and smallholder farming in Africa: the heretics' view. *Field Crops Research* 114: 23–34.

Giller, K.E., Corbeels, M., Nyamangara, J., Triomphe, B., Affholder, F., Scopel, E. and Tittonell, P. 2011. A research agenda to explore the role of conservation agriculture in African smallholder farming systems. *Field Crops Research* 124(3): 468–472.

Gladwin, C.H and Butler, J. 1982. Gardening: a survival strategy for the small, part-time Florida farm. *Proceedings of the Florida State Horticultural Society* 95: 264–268.

Gliessman, S.R. 2005. Agroecology and agroecosystems. In: Pretty, J.N. (ed.) *The Earthscan Reader in Sustainable Agriculture*. London: Earthscan.

Gliessman, S.R. 2014. *Agroecology: Ecological Processes in Sustainable Agriculture*. Third Edition. Boca Raton, FL: CRC Press. Global Green Growth Institute. 2012. URL: www.gggi.org

Glover, J.D. 2011. The system of rice intensification: time for an empirical turn. *NJAS – Wageningen Journal of Life Sciences* 57(3–4): 217–224.

Glover, J.D. and Reganold, J.P. 2010. Perennial grains: food security for the future. *Issues in Science and Technology* 26(2): 41–47.

Godfray, H.C.J., Beddington, J.R., Crute, I.R., Haddad, L., Lawrence, D., Muir, J.F., *et al.* 2010. Food security: the challenge of feeding 9 billion people. *Science* 327: 812–818.

Goldschmidt, W. 1978 [1946]. *As You Sow: Three Studies into the Social Consequences of Agribusiness*. Montclair, NJ: Allanheld, Osmunn.

Gomez-Pompa, A. and Kaus, A. 1992. Taming the wilderness myth. *Bioscience* 42(4): 271–279.

Grameen Bank. 2017. URL: www.grameen.com

Gray, R. and Dayananda, B. 2014. Structure of public research. In: Smyth, S., Phillips, P.W.B. and Castle, D. (eds) *Handbook on Agriculture, Biotechnology and Development*. Cheltenham: Edgar Elgar.

Groenfeldt, D. and Sun, P. 1997. Demand management of irrigation systems through users' participation. In: Kay, M., Franks, T. and Smith, L. (eds) *Water: Economics, Management, and Demand* (pp.304–312). London: E & FN Spon.

Grovermann, C., Schreinemachers, P. and Berger, T. 2013. Quantifying pesticide overuse from farmer and societal points of view: an application to Thailand. *Crop Protection* 53: 161–168.

Gunton, R.M., Firbank, L.G., Inman, A. and Winter, D.M. 2016. How scalable is sustainable intensification? *Nature Plants* 2: 1–4.

Gurr, G.M., Lu, Z., Zheng, X., Xu, H., Zhu, P., Chen, G., *et al.* 2016. Multi-country evidence that crop diversification promotes ecological intensification of agriculture. *Nature Plants* 2: 16014.

Hall, J. 2008. The Role of Social Capital in Farmers' Transition to more Sustainable Land Management. PhD Thesis. University of Essex.

Hall, J. and Pretty, J.N. 2008. Then and now: Norfolk farmers' changing relationships and linkages with government agencies during transformations in land management. *Journal of Farm Management* 13(6): 393–418.

Hall, N.M., Bocary, K., Janet, D., Ute, S., Amadou, N. and Tobo, R. 2006. Effect of improved fallow on crop productivity, soil fertility and climate-forcing gas emissions in semi-arid conditions. *Biology and Fertility of Soils* 42: 224–230.

Halwart, M. 2013. Valuing aquatic biodiversity in agricultural landscapes. In: Fanzo, J., Hunter, D., Borelli, T. and Mattei, F. (eds) *Diversifying Food and Diets – Using Agricultural Biodiversity to Improve Nutrition and Health* (pp. 88–108). London: Earthscan.

Hardman, C.J., Harrison, D.P., Shaw, P.J., Nevard, T.D., Hughes, B., Potts, S.G. and Norris, K. 2016. Supporting local diversity of habitats and species on farmland: a comparison of three wildlife-friendly schemes. *Journal of Applied Ecology* 53(1): 171–180.

Hartley, S. 2017. Agroecological approaches to sustainable intensification. *International Journal of Agricultural Sustainability* (in press).

He, J., Kuhn, N.J., Zhang, X.M., Zhang, X.R., & Li, H.W. 2009. Effects of 10 years of conservation tillage on soil properties and productivity in the farming-pastoral ecotone of Inner Mongolia, China. *Soil Use Manage* 25: 201–209.

He, J., Li, H., Rasaily, R.G., Wang, Q., Cai, G., Su, Y., *et al.* 2011. Soil properties and crop yields after 11 years of no tillage farming in wheat–maize cropping system in North China Plain. *Soil and Tillage Research* 113(1): 48–54.

Head, G. and Savinelli, C. 2008. Adapting insect resistance management programs to local needs. In: Onstad, D. (ed.) *Insect Resistance Management: Biology, Economics and Prediction.* Amsterdam: Elsevier.

Heap, I. 2014. Global perspective on herbicide-resistant weeds. *Pest Management Science* doi: 10.1002/ps.3696

Hensler, A., Barker, D., Sulc, S., Loerch, S. and Owens, L.B. 2007. Management Intensive Grazing and Continuous Grazing of Hill Pasture by Beef Cattle. Coshocton, OH: USDA Agricultural Research Service.

Herren, H.R., Neuenschwandrer, P., Hennessey, R.D. and Hammond, W.N.O. 1987. Introduction and dispersal of *Epidinocarsis lopezi* (Hym., Encyrtidae), an exotic parasitoid of the cassava mealybug, *Phenacoccus manihoti* (Hom., Pseudococcidae), in Africa. *Agriculture, Ecosystems & Environment* 19(2): 131–144.

Hill, S. 1985. Redesigning the food system for sustainability. *Alternatives* 12: 32–36.

Hill, S. 2014. Considerations for enabling the ecological redesign of organic and conventional agriculture: a social ecology and psychological perspective. In: Bellon, S. and Penvern, S. (eds) *Organic Farming, Prototype for Sustainable Agriculture.* Dordrecht: Springer.

Hill, S. 2015. Personal priorities for organics to realise its potential. *Journal of Organic Systems* 10: 1–2.

Himmelstein, J., Ares, A., Gallagher, D. and Myers, J. 2017. A meta-analysis of intercropping in Africa: impacts on crop yield, farmer income, and integrated pest management effects. *International Journal of Agricultural Sustainability* 15(1): 1–10.

Hoi, P.V., Mol, A.P., Oosterveer, P., van den Brink, P.J. and Huong, P.T. 2016. Pesticide use in Vietnamese vegetable production: a 10-year study. *International Journal of Agricultural Sustainability* 14(3): 325–338.

Holdren, J. and Ehrlich, P. 1974. Human population and the global environment. *American Scientist* 62: 282–292.

Holmann, F., Rivas, L., Argel, P. and Pérez, E. 2004. Impact of the adoption of *Brachiaria* grasses: Central America and Mexico. *Livestock Research for Rural Development* 16(12): 2004.

Holt-Giménez, E. and Altieri, M.A. 2013. Agroecology, food sovereignty, and the new green revolution. *Agroecology and Sustainable Food Systems* 37: 90–102.

Horlings, L.G. and Marsden, T.K. 2011. Towards the real green revolution? Exploring the conceptual dimensions of a new ecological modernisation of agriculture that could 'feed the world'. *Global Environmental Change* 21: 441–452.

Hu, L., Zhang, J., Ren, W., Guo, L., Cheng, Y., Li, J., *et al.* 2016. Can the co-cultivation of rice and fish help sustain rice production? *Scientific Reports* 6.

Huang, J. and Ding, J. 2016. Institutional innovation and policy support to facilitate small-scale farming transformation in China. *Agricultural Economics* 47(S1): 227–237.

IAASTD. 2009. *Agriculture at a Crossroads, International Assessment of Agricultural Knowledge, Science and Technology for Development.* Washington, DC: Island Press.

Iannotti, L., Cunningham, K. and Ruel, M. 2009. Improving Diet Quality and Micronutrient Nutrition: Homestead Food Production in Bangladesh. IFPRI Discussion Paper 00928. Washington, DC: International Food Policy Research Institute.

ICIPE. 2013. Stories of our success: positive outcomes from push–pull farming systems. www.push-pull.net/farmers_success.pdf

Ickowitz, A., Powell, B., Salim, M.A. and Sunderland, T.C.H. 2013. Dietary quality and tree cover in Africa. *Global Environmental Change* 24: 287–294.

IFAD and UNEP. 2013. Smallholders, Food Security and the Environment. IFAD. https://www.ifad.org/documents/10180/666cac24-14b6-43c2-876d-9c2d1f01d5dd

IFOAM–EU. 2017. Organic in Europe. URL: www.ifoam-eu.org/en/organic-europe IFPRI (International Food Policy Research Institute). 2016. Global Food Policy Report. Washington, DC: IFPRI.

Institut pour Agriculture Durable (IAD). 2011. Agriculture 2050 Starts Here and Now. Paris: IAD.

IPCC. 2007. Climate Change 2007: Impacts, Adaptation and Vulnerability. Contribution of Working Group II to the Fourth Assessment Report of the Intergovernmental Panel on Climate Change.

IPES-Food. 2016. From uniformity to diversity: a paradigm shift from industrial agriculture to diversified agroecological systems. International Panel of Experts on Sustainable Food Systems. www.ipes-food.org

IPES-Food. 2017. Unravelling the food–health nexus. Addressing practices, political economy and power relations to build healthier food systems. URL: www.ipes-food.org/new-report-an-overwhelming-case-for-action-expert-panel-identifies-unacceptable-toll-of-food-and-farming-systems-on-human-health

ISAAA. 2016. Global Status of Commercialized Biotech/GM Crops: 2013. Brief 52–2017. http://isaaa.org/resources/publications/briefs/52/executivesummary/default.asp

IUCN. 2013. Food Security Policies: Making the Ecosystem Connections. Gland, Switzerland: IUCN.

Jackson, T. 2017. *Prosperity without Growth. Foundations for the Economy of Tomorrow.* Second Edition. London: Routledge.

Jacobsen, S., Sørensen, M., Pedersen, S.M. and Weiner, J. 2013. Feeding the world: genetically modified crops versus agricultural biodiversity. *Agronomy for Sustainable Development* doi: 10.1007/s13593-013-0138-9.

Jat, R., Kanwar, S. and Kassam, A. 2014. *Conservation Agriculture: Global Prospects and Challenges.* Wallingford: CABI.

Jatoe, J.P.D., Lankoandè, D.G. and Sumberg, J. 2015. Does sustainable agricultural growth require a system of innovation? Evidence from Ghana and Burkina Faso. *International Journal of Agricultural Sustainability* 13(2): 104–119.

Jepson, P.C., Guzy, M., Blaustein, K., Sow, M., Sarr, M., Mineau, P. and Kegley, S. 2014. Measuring pesticide ecological and health risks in West African agriculture to establish an enabling environment for sustainable intensification. *Philosophical Transactions of the Royal Society B* 369. doi: 10.1098/rstb.2013.0491

Jones, D.L., Cross, P., Withers, P.J.A., DeLuca, T.H., Robinson, D.A., Quilliam, *et al.* 2013. Nutrient-stripping: the global disparity between food security and soil nutrient stocks. *Journal of Applied Ecology* 50: 851–862.

Ju, X., Gu, B., Wu, Y. and Galloway, J.N. 2016. Reducing China's fertilizer use by increasing farm size. *Global Environmental Change* 41: 26–32.

Karabayev, M. 2012. Conservation agriculture adoption in Kazakhstan. Presentation to the WIPO Conference on Innovation and Climate Change, 11–12 July 2011. Geneva.

Karabayev, M. and Suleimenov, M. 2010. Adoption of conservation agriculture in Kazakhstan. In: Lead papers 4th World Congress on Conservation Agriculture: Innovations for Improving Efficiency, Equity and Environment. 4–7 February 2009. New Delhi.

Karabayev, M., Morgounov, A., Braun, H-J., Wall, P., Sayre, K., Zelenskiy, Y., *et al.* 2014. Effective approaches to wheat improvement in Kazakhstan: breeding and conservation agriculture. *Journal of Bahri Dagdas Crop Research* (1–2): 50–53.

Kartaatmadja, S., Pane, H., Wirajaswadi, L., Sembiring, H., Simatupang, S., Bachrein, S., *et al.* 2004. Optimising use of natural resources and increasing rice productivity. In: *Conserving Soil and Water for Society: Sharing Solutions*. ISCO 2004–13th International Soil Conservation Organisation Conference – Brisbane, Australia, July.

Kassam, A., Friedrich, T., Shaxson, F. and Pretty, J.N. 2009. The spread of conservation agriculture: justification, spread and uptake. *International Journal of Agricultural Sustainability* 7(4): 292–320.

Kassam, A., Friedrich, T., Shaxson, F., Bartz, H., Mello, I., Kienzle, J. and Pretty, J.N. 2014. The spread of Conservation Agriculture: policy and institutional support for adoption and uptake. *Field Actions Science Reports* 7. http://factsreports.revues.org/3720

Kell, D.B. 2011. Breeding crop plants with deep roots: their role in sustainable carbon, nutrient and water sequestration. *Annals of Botany* 108: 407–418.

Kelly, R.L. 1995. *The Foraging Spectrum: Diversity in Hunter-Gatherer Lifeways*. Washington, DC: Smithsonian.

Kenmore, P., Carino, F., Perez, G. and Dyck, V. 1984. Population regulation of the rice brown plant hopper Nilaparvata lugens (Stal) within rice fields in the Philippines. *Journal of Plant Protection in the Tropics* 1: 19–38.

Kent, S. (ed.) 1989. *Farmers as Hunters: The Implications of Sedentism*. Cambridge: Cambridge University Press.

Ketelaar, J.W. and Abubakar, A.L. 2012. Sustainable intensification of agricultural production. In: Kim, M. and Diong, C.H. (eds) *Biology Education for Social and Sustainable Development* (pp. 173–181). Rotterdam: Sense Springer.

Ketelaar, J.W., Abubakar, A.L.M., Van Du, P., Widyastama, C., Phasouysaingam, A., Binamira, J. and Dung, N.T. 2018. Save and grow: translating policy advice into field action for sustainable intensification of rice production. In: Nagothu, U.S. (ed.) *Agricultural Development and Sustainable Intensification: Technology and Policy Challenges in the Face of Climate Change*. London: Routledge.

Khan, Z.R., Midega, C.A.O., Pittchar, J., Pickett, J.A. and Bruce, T. 2011. Push–pull technology: a conservation agriculture approach for integrated management of insect pests, weeds and soil health in Africa. *International Journal of Agricultural Sustainability* 9: 162–170.

Khan, Z.R., Midega, C.A.O., Pittchar, J.O., Murage, A.W., Birkett, M.A., Bruce, T.J.A. and Pickett, J.A. 2014. Achieving food security for one million sub-Saharan African poor through push–pull innovation by 2020. *Philosophical Transactions of the Royal Society (B)* 369(1639): 201.

Khan, Z.R., Midega, C., Pittchar, J.O., Murage, A. and Pickett, J. 2017. Climate-smart push–pull: a conservation agriculture technology for food security and environmental sustainability in Africa. Wallingford: CABI.

King, F.H. 1911. *Farmers of Forty Centuries*. Emmaus, PA: Rodale Press.

Klink, C.A. and Moreira, A.G. 2002. Past and current human occupation, and land use. In: Oliveira, P. and Marquis, R. (eds) *The Cerrados of Brazil: Ecology and Natural History of a Neotropical Savanna* (pp.69–88). New York: Columbia University Press.

Kluthcouski, J. and Pacheco-Yokoyama, L.P. 2006. Crop–Livestock Integration Options. URL: www.fao.org/ag/agp/agpc/doc/integration/papers/integration_options.htm

Kluthcouski, J., Cobucci, T., Aidar, H., Yokoyama, L.P., Oliveira, I.P., Costa, J.L.S., *et al.* 2000. Integração lavoura /pecuária pelo consórcio de culturas anuais com forrageiras, em áreas de lavoura, nos sistemas direto e convencional, Embrapa Arroz e Feijâo Circular Técnica No. 38.

Knutson, R., Penn, J. and Flinchbaugh, B. 1998. *Agricultural and Food Policy*. Fourth Edition. Upper Saddle River, NJ: Prentice Hall.

Koepf, H.H. 1989. *The Biodynamic Farm*. Hudson, NY: Anthroposophic Press.

Koh, I., Lonsdorf, E.V., Williams, N.M., Brittain, C., Isaacs, R., Gibbs, J. and Ricketts, T. H. 2016. Modelling the status, trends, and impacts of wild bee abundance in the United States. *PNAS* 113: 140–145.

Kohli, M.M. and Fraschina, J. 2009. Adapting wheats to zero tillage in maize–wheat–soybean rotation system. 4th World Congress of World Agriculture. URL: www.fao.org/ag/Ca/doc/wwcca-leadpapers.pdf#page=218

Koleva, N.G. and Schneider, U.A. 2009. The impact of climate change on the external cost of pesticide application in US agriculture. *International Journal of Agricultural Sustainability* 7: 203–216.

Koohafkan, P., Altieri, M. and Holt-Giménez, E. 2012. Green agriculture: foundations for biodiverse, resilient and productive agricultural systems. *International Journal of Agricultural Sustainability* 10(1): 61–75.

Krupnik, T.J., Shennan, C., Settle, W., Demont, M., Ndiaye, A.B. and Rodenburg, J. 2012. Improving irrigated rice production in the Senegal River Valley through experiential learning and innovation. *Agricultural Systems* 109: 101–112.

Kuhn, T. [1962] 1970. *The Structure of Scientific Revolutions*. Second Edition. Chicago, IL: Chicago University Press.

Kulkarni, S.A. and Tyagi, A.C. 2012. Participatory irrigation management: understanding the role of cooperative culture. International Commission on Irrigation and Drainage.

Kulkarni, S.C., Levin-Rector, A., Ezzati, M. and Murray, C.J.L. 2011. Falling behind: life expectancy in US counties from 2000 to 2007 in an international context. *Population Health Metrics* 9: 16.

Kwate, N.O.A., Yau, C., Loh, J. and Williams, D. 2009. Inequality in obesigenic environments: fast food density in New York City. *Health & Place* 15(1): 364–373.

Lal, R. 2004. Agricultural activities and the global carbon cycle. *Nutrient Cycling in Agroecosystems* 70: 103–116.

Lal, R. 2014. Abating climate change and feeding the world through soil carbon sequestration. In: Dent, D. (ed.) *Soil as World Heritage*. Dordrecht: Springer.

Lamberth, C., Jeanmart, S., Luksch, T. and Plant, A. 2013. Current challenges and trends in the discovery of agrochemicals. *Science* 341: 742–746.

Lamine, C. 2011. Transition pathways towards a robust ecologization of agriculture and the need for system redesign. Cases from organic farming and IPM. *Journal of Rural Studies* 27(2): 209–219.

Lampkin, N.H., Pearce, B.D., Leake, A.R., Creissen, H., Gerrard, C.L., Girling, R., *et al*. 2015. The role of agroecology in sustainable intensification. Report for the Land Use Policy Group. Organic Research Centre, Elm Farm and Game and Wildlife Conservation Trust.

Lang, T. and Barling, D. 2012. Food security and food sustainability: reformulating the debate. *Geographical Journal* 178(4): 313–326.

Larwanou, M. and Adam, T. 2008. Impacts de la re'ge'ne'ration naturelle assiste'e au Niger: Etude de quelques cas dans les Re'gions de Maradi et Zinder. Synthe`se de 11 me'moires d'e'tudiants de 3e`me cycle de l'Universite' AbdouMoumouni de Niamey, Niger.

Lathrap, D.W. 1968. The hunting economics of the tropical forest zone of South America. In: Lee, R.B. and DeVore, I. (eds) *Man the Hunter*. Chicago, IL: Aldine.

Leach, A.W. and Mumford, J.D. 2008. Pesticide environmental accounting: a method for assessing the external costs of individual pesticide applications. *Environmental Pollution* 151: 139–147.

Leach, A.W. and Mumford, J.D. 2011. Pesticide environmental accounting: a decision-making tool estimating external costs of pesticides. *J. Verbr. Lebensm.* (Journal of Consumer Protection and Food Safety) 6 (Suppl 1): S21–S26.

Lechenet, M., Dessaint, F., Py, G., Makowski, D. and Munier-Jolain, N. 2017. Reducing pesticide use while preserving crop productivity and profitability on arable farms. *Nature Plants* 3: 17008.

Lee, J.S.H., Abood, S., Ghazoul, J., Barus, B., Obidzinski, K. and Koh, L.P. 2014. Environmental impacts of large-scale oil palm enterprises exceed that of smallholdings in Indonesia. *Conservation Letters* 7(1): 25–33.

Lee, R.B. and DeVore, I. (eds). 1968. *Man the Hunter.* Chicago, IL: Aldine.

Lee-Smith, D. 2010. Cities feeding people: an update on urban agriculture in equatorial Africa. *Environment and Urbanization* 22: 483–499.

Leeuwis, C. and Van den Ban, A. 2004. *Communication for Rural Innovation: Rethinking Agricultural Extension.* Third Edition. Oxford: Blackwell and CTA.

Letourneau, D.K., Armbrecht, I., Rivera, B.S., Lerma, J.M., Carmona, E.J., Daza, M.C., *et al.* 2011. Does plant diversity benefit agroecosystems? A synthetic review. *Ecological Applications* 21: 9–21.

Li, H.W., Gao, H.W., Wu, H. D., Li, W.Y., Wang, X.Y. and He, J. 2007. Effects of 15 years of conservation tillage on soil structure and productivity of wheat cultivation in northern China. *Australian Journal of Soil Research* 45: 344–350.

Li Hongwen, He Jin, Bharucha, Z.P., Lal, R. and Pretty, J. 2016. Improving China's food and environmental security with conservation agriculture. *International Journal of Agricultural Sustainability* 14: 377–391.

Li Wenhua. 2001. *Agro-Ecological Farming Systems in China.* Man and the Biosphere Series Volume 26. Paris: UNESCO.

Liang, K., Yang, T., Zhang, S., Zhang, J., Luo, M., Fu, L. and Zhao, B. 2016. Effects of intercropping rice and water spinach on net yields and pest control: an experiment in southern China. *International Journal of Agricultural Sustainability* 14(4): 448–465.

Lin, X.Q., Zhu, D.F., Chen, H.Z., Cheng, S.H. and Uphoff, N. 2009. Effect of plant density and nitrogen fertilizer rates on grain yield and nitrogen uptake of hybrid rice (*Oryza sativa* L.). *Journal of Agricultural Biotechnology and Sustainable Development* 1: 44–53.

Little, R., Maye, D. and Ilbery, B. 2010. Collective purchase: moving local and organic foods beyond the niche market. *Environment and Planning A* 42(8): 1797–1813.

Liu, J. and Innes, J.L. 2015. Participatory forest management in China: key challenges and ways forward. *International Forestry Review* 17(2): 1–8.

Liu, S.S., Rao, A. and Vinson, S.B. 2014. Biological control in China: past, present and future – an introduction to this special issue. *Biological Control* 68: 1–5.

Lobao, L. 1990. *Locality and Inequality: Farm and Industry Structure and Socio-Economic Conditions.* New York: State University of New York Press.

Lorenz, K. and Lal, R. 2014. Soil organic carbon sequestration in agroforestry systems: a review. *Agronomy for Sustainable Development* 34(2): 443–454.

Lowder, S.K., Skoet, J. and Raney, T. 2016. The number, size, and distribution of farms, small-holder farms, and family farms worldwide. *World Development* 87: 16–29 https://doi.org/10.1016/j.worlddev.2015.10.041

Lu, Y., Chadwick, D., Norse, D., Powlson, D. and Shi, W. 2015. Sustainable intensification of China's agriculture: the key role of nutrient management and climate change mitigation and adaptation. *Agriculture, Ecosystems and Environment* 209: 1–4.

McBratney, A., Field, D.J. and Koch, A. 2014. The dimensions of soil security. *Geoderma* 213: 203–213.

McDonald, A.J., Hobbs, P.R. and Riha, S.J. 2006. Does the system of rice intensification outperform conventional best management? A synopsis of the empirical record, *Field Crops Research* 96(1): 31–36.

McDonald, A.J., Hobbs, P.R. and Riha, S.J. 2008. Stubborn facts: still no evidence that the system of rice intensification out-yields best management practices (BMPs) beyond Madagascar. *Field Crops Research* 108(2): 188–191.

McKee, R. 1998. *Story. Substance, Structure, Style and the Principles of Screenwriting*. New York: Harper Collins.

MacMillan, T. and Benton, T.G. 2014. Engage farmers in research. *Nature* 509(7498): 25.

MacRae, R.J., Henning, J. and Hill, S.B. 1993. Strategies to overcome barriers to the development of sustainable agriculture in Canada: the role of agribusiness. *Journal of Agricultural & Environmental Ethics* 6(1): 21–51.

Macy, J. 2005. The Great Turning. Available online: https://humanisticpaganism.com/2015/04/04/the-great-turning-by-joanna-macy/

MacWilliam, S., Wismer, M. and Kulshreshtha, S. 2014. Life cycle and economic assessment of Western Canadian pulse systems: the inclusion of pulses in crop rotations. *Agricultural Systems* 123: 43–53.

Mahon, N., Crute, I., Simmons, E. and Islam, M.M. 2017. Sustainable intensification – 'oxymoron' or 'third-way'? A systematic review. *Ecological Indicators* 74: 73–97. doi: 10.1016/j.ecolind.2016.11.001

Mangan, J. and Mangan, M.S.A. 1998. Comparison of two IPM training strategies in China: the importance of concepts of the rice ecosystem for sustainable insect pest management. *Agriculture and Human Values* 15: 209–221.

Manu-Aduening, J.A., Peprah, B., Bolfrey-Arku, G. and Aubyn, A. 2014. Promoting farmer participation in client-oriented breeding: lessons from participatory breeding for farmer-preferred cassava varieties in Ghana. *Advanced Journal of Agricultural Research* 2(002): 008–017.

Marongwe, L.S., Kwazira, K., Jenrich, M., Thierfelder, C., Kassam, A. and Friedrich, T. 2011. An African success: the case of conservation agriculture in Zimbabwe. *International Journal of Agricultural Sustainability* 9(1): 153–161.

Marsden, T. 2014. Conclusions: building the food sustainability paradigm: research needs, complexities, opportunities. In: Marsden, T. and Morley, A. (eds) *Sustainable Food Systems: Building a New Paradigm*. Abingdon: Routledge.

Maxwell, D. 2012. Food security and its implications for political stability: a humanitarian perspective. Paper prepared for workshop on Food Security and Its Implications for Global Stability. Cornell University, Ithaca, NY, 18–19 June. www.fao.org/fileadmin/templates/cfs_high_level_forum/documents/FS-Implications-Political_Stability-Maxwell.pdf (31 August 2014).

Maxwell, S.L., Fuller, R.A., Brooks, T.M. and Watson, J.E.M. 2016. Biodiversity: the ravages of guns, nets and bulldozers. *Nature* 536: 143–145.

MEA (Millennium Ecosystem Assessment). 2005. *Ecosystems and Well-being.* Washington, DC: Island Press.

Meadowcroft, J. 2011. Engaging with the *politics* of sustainability transitions. *Environmental Innovation and Societal Transitions* 1(1): 70–75.

Meadows, D.H., Meadows, D.L., Randers, J. and Behrens, W.W. 1972. *The Limits to Growth.* New York: Universe Books.

Meggers, B.J. 1954 Environmental limitation on the development of culture. *American Anthropologist* 56: 801–824.

Méndez, V.E., Lok, R. and Somarriba, E. 2001. Interdisciplinary analysis of homegardens in Nicaragua: micro-zonation, plant use and socioeconomic importance. *Agroforestry Systems* 51: 85–96.

MFSC (Ministry of Forest and Soil Conservation, Nepal). 2013. *Persistence and Change: Review of 30 Years of Community Forestry in Nepal.* Kathmandu. E-book (at www.msfp.org.np/uploads/publications/file/ebook_interactiv_20130517095926.pdf).

Milder, J.C., Garbach, K., DeClerck, F.A.J., Driscoll, L., Montenegro, M. 2012. An Assessment of the Multi-Functionality of Agroecological Intensification. A report prepared for the Bill and Melinda Gates Foundation.

Miller, J.W. and Atanda, T. 2011. The rise of peri-urban aquaculture in Nigeria. *International Journal of Agricultural Sustainability* 9(1): 274–281.

Mittra, S., Sugam, R. and Ghosh, A. 2014. Collective action for water security and sustainability. Preliminary investigations. Council on Energy, Environment and Water. URL: http://ceew.in/pdf/CEEW-2030-WRG-Collective-Action-for-Water-Security-and-Sustainability-Report-19Aug14.pdf

MOEP (Ministry of Environmental Protection). 2014. *Bulletin of Chinese Environmental Status.* Beijing, China.

Moffatt, I. 2000. Ecological footprints and sustainable development. *Ecological Economics* 32: 359–362.

Mollison, B. 1988. *Permaculture: A Designer's Manual.* Tyalgum, NSW: Tagari Publications.

Montpellier Panel. 2013. Sustainable Intensification: A New Paradigm for African Agriculture. London: Agriculture for Impact.

Morgan, A. 1999. *Eat the Big Fish.* London: John Wiley.

Morgan, J.A., Follet, R.F., Allen, L.H., Del Grosso, S., Derner, J.D., Dijkstra, F., *et al.* 2010. Carbon sequestration in agricultural lands of the United States. *Journal of Soil and Water Conservation* 65(1): 7A–13A.

Moss, B. 2008. Water pollution by agriculture. *Philosophical Transactions of the Royal Society (B)* 363(1491): 659–666.

MOWR (Ministry of Water Resources). 2013. *Bulletin of Chinese Soil and Water Conservation.* Beijing, China.

Muhanji, G., Roothaert, R.L., Webo, C. and Stanley, M. 2011. African indigenous vegetable enterprises and market access for small-scale farmers in East Africa. *International Journal of Agricultural Sustainability* 9(1): 194–202.

Murthy, I.K., Gupta, M., Tomar, S., Munsi, M., Tiwari, R., Hedge, G.T. and Ravindranath, N.H. 2013. Carbon sequestration potential of agroforestry systems in India. *Earth Science & Climate Change* 4: 1 http://dx.doi.org/10.4172/2157-7617.1000131

Mwanga, R.O. and Ssemakula, G. 2011. Orange-fleshed sweet potatoes for food, health and wealth in Uganda. *International Journal of Agricultural Sustainability* 9(1): 42–49.

Myrick, S., Norton, G., Selvaraj, K.N., Natarajan, K. and Muniappan, R. 2014. Economic impact of classical biological control of papaya mealybug in India. *Crop Protection* 56: 82–86.

Naess, A. 1973. The shallow and the deep, long-range ecology movement. *Inquiry* 16: 95–100.

Nagothu, U.S. (ed.) 2018. *Agricultural Development and Sustainable Intensification: Technology and Policy Challenges in the Face of Climate Change.* London: Routledge.

Nath, T.K., Jashimuddin, M., Hasan, M.K., Shahjahan, M. and Pretty, J. 2016. The sustainable intensification of agroforestry in shifting cultivation areas of Bangladesh. *Agroforestry Systems* 90(3): 405–416.

National Farm to School Network. 2013. Nourishing kids and community. URL: www.farmtoschool.org/aboutus.php

Natural Capital Committee. 2017. Economic Valuation and Its Applications in Natural Capital Management and the Government's 25 Year Environment Plan. NCC.

Naylor, R.L., Falcon, W.P., Goodman, R.M., Jahn, M.M., Sengooba, T., Tefera, H. and Nelson, R.J. 2004. Biotechnology in the developing world: a case for increased investments in orphan crops. *Food Policy* 29(1): 15–44.

NCGA (National Corn Growers Association). 2014. *World of corn*. URL: www.ncga.com/upload/files/%20documents/pdf/woc-2014.pdf

NEA. 2011. Synthesis of the Key Findings. Report from the UK National Ecosystem Assessment. Cambridge: UNEP–WCMC.

NEF (New Economics Foundation). 2003. New survey launched at localism conference. Press Release, 16 May.

NEF. 2012. The Happy Planet Index: 2012 Report. A Global Index of Sustainable Well-being. URL: https://static1.squarespace.com/static/5735c421e321402778ee0ce9/t/578cb7e8b3db2b247150c93e/1468839917409/happy-planet-index-report-2012.pdf

Neuenschwander, P. 2001. Biological control of the cassava mealybug in Africa: a review. *Biological Control* 21(3): 214–229.

Nga, N., Rodriguez, D., Son, T. and Buresh, R.J. 2010. Development and impact of site-specific nutrient management in the Red River Delta of Vietnam (pp. 317–334). In: Palis, F.G., Singleton, G.R., Casimero, M.C. and Hardy, B. (eds) *Research to Impact. Case Studies for Natural Resource Management for Irrigated Rice in Asia*. International Rice Research Institute. Los Baños, Philippines.

Nightingale, A. and Sharma, J.R. 2014. Conflict resilience among community forestry user groups: experiences in Nepal. *Disasters* 38(3): 517–539.

Niñez, V.K. 1984. Household gardens: theoretical considerations on an old survival strategy. Potatoes in Food Systems Research Series Report No. 1. International Potato Center. http://pdf.usaid.gov/pdf_docs/PNAAS307.pdf (31 August 2014).

Niñez, V.K. 1985. Working at half-potential: constructive analysis of homegarden programme in the Lima slums with suggestions for an alternative. http://archive.unu.edu/unupress/food/8F073e/8F073E02.htm

Norse, D. 2012. Low carbon agriculture: objectives and policy pathways. *Environmental Development* 1(1): 25–39.

Norse, D. and Ju, X. 2015. Environmental costs of China's food security. *Agriculture, Ecosystems and Environment* 209: 5–14.

Norse, D., Li, J., Leshan, J. and Zheng, Z. 2001. *Environmental Costs of Rice Production in China*. Bethesda, MD: Aileen Press.

NRC (National Research Council). 1989. *Alternative Agriculture*. Washington, DC: National Academies Press.

NRC. 2010. *Towards Sustainable Agricultural Systems in the 21st Century*. Washington, DC: National Academies Press.

Nurbekov, A., Akramkhanov, A., Lamers, J., Kassam, A., Friedrich, T., Gupta, R., *et al.* 2014. Conservation agriculture in Central Asia. In: Jat, R., Sahrawat, K. and Kassam, A. (eds) *Conservation Agriculture: Global Prospects and Challenge*. Wallingford: CABI.

O'Bannon, C., Carr, J., Seekell, D.A. and D'Odorico, P. 2014. Globalization of agricultural pollution due to international trade. *Hydrology and Earth System Sciences* 18(2): 503–510.

OECD. 1996. *The Knowledge-based Economy*. Paris: OECD.

OECD. 2011. *Towards Green Growth*. Paris: OECD.

OECD. 2013. *Dataset: 2013 Edition of the OECD Environmental Database*. Paris: OECD.

OECD. 2017. Obesity Update 2017. URL: www.oecd.org/health/obesity-update.htm

Ongachi, W., Onwonga, R., Nyanganga, H. and Okry, F. 2017. Comparative analysis of video mediated learning and farmer–field school approach on adoption of *Striga* control technologies in Western Kenya. *International Journal of Agricultural Extension* 5(1): 01–10.

Oreskes, N. and Conway, E.M. 2010. Defeating the merchants of doubt. *Nature* 465: 686–687.

Ostrom, E. 1990. *Governing the Commons: The Evolution of Institutions for Collective Action*. Cambridge: Cambridge University Press.

Owenya, M.Z., Mariki, W.L., Kienzle, J., Friedrich, T. and Kassam, A. 2011. Conservation agriculture (CA) in Tanzania: the case of the Mwangaza B CA farmer field school (FFS), Rhotia Village, Karatu District, Arusha. *International Journal of Agricultural Sustainability* 9(1): 145–152.

Pacheco, A.R., de Queiroz Chaves, R. and Lana Nicoli, C.M. 2013. Integration of crops, livestock, and forestry: a system of production for the Brazilian Cerrados. In: Hershey, C.H. and Neate, P. (eds) *Eco-efficiency: From Vision to Reality* (pp. 51–60). Cali, Colombia: Centro Internacional de Agricultura Tropical (CIAT).

Pahl-Wostl, C. 2009. A conceptual framework for analysing adaptive capacity and multi-level learning processes in resource governance frameworks. *Global Environmental Change* 19(3): 354–365.

Palanisami, K., Karunakaran, K.R., Amarasinghe, U. and Ranganathan, C.R. 2013. Doing different things or doing it differently: rice intensification practices in 13 states of India. *Economic and Political Weekly* 48(8): 51–67.

Palis, F.G. 2006. The role of culture in farmer learning and technology adoption: a case study of farmer field schools among rice farmers in central Luzon, Philippines. *Agriculture and Human Values* 23(4). doi: 10.1007/s10460-006-9012-6

Pan, G., Zhou, P., Li, Z., Smith, P., Li, L., Qiu, D., Zhang, X., Xu, X., Shen, S. and Chen, X. 2009. Combined inorganic/organic fertilization enhances N efficiency and increases rice productivity through organic carbon accumulation in a rice paddy from the Tai Lake region, China. *Agriculture, Ecosystems & Environment* 131(3): 274–280.

Papademetriou, M.K. 2000. Rice production in the Asia-Pacific region. Issues and perspectives. In: Papademetriou, M.K, Dent, F.J. and Herath, E.M. (eds) *Bridging the Rice Yield Gap in the Asia-Pacific Region*. Rome: FAO.

Parsa, S., Morse, S., Bonifacio, A., Chancellor, T.C.B., Condori, B., Crespo-Perez, V *et al.* 2014. Obstacles to integrated pest management adoption in developing countries. *PNAS* 111(10): 3889–3894.

Pearce, D. and Tinch, R. 1998. The true price of pesticides. In: Vorley, W. and Keeney, D. (eds) *Bugs in the System: Redesigning the Pesticide Industry for Sustainable Agriculture* (pp. 50–93). London: Earthscan.

Peer, N., Kengne, A., Motala, A.A. and Mbanya, J.C. 2014. Diabetes in the Africa region: an update. *Diabetes Research and Clinical Practice* 103(2): 197–205.

Perelman, M. 1976. Efficiency in agriculture: the economics of energy. In: Merril, R. (ed.) *Radical Agriculture*. New York: Harper and Row.

Petersen, P., Tardin, J.M. and Marochi, F. 2000. Participatory development of non-tillage systems without herbicides for family farming: the experience of the center-south region of Parana. *Environment, Development and Sustainability* 1: 235–252.

PHE (Public Health England). 2013. Obesity and the Environment: Increasing Physical Activity and Active Travel. London: Public Health England.

Phillips, D., Waddington, H. and White, H. 2014. Better targeting of farmers as a channel for poverty reduction: a systematic review of farmer field schools targeting. *Development Studies Research* 1(1): 113–136.

Pilgrim, S. and Pretty, J.N. (eds) 2010. *Nature and Culture: Rebuilding Lost Connections.* London: Earthscan.

Pingali, P.L. and Roger, P.A. 1995. *Impact of Pesticides on Farmers' Health and the Rice Environment.* Dordrecht: Kluwer Academic.

Pingali, P. and Raney, T. 2005. From the Green Revolution to the Gene Revolution. How Will the Poor Fare? ESA Working Paper No. 05–09. November. Agricultural and Economics Division, FAO. www.fao.org/3/a-af276t.pdf

Poffenberger, M. and Zurbuchen, M.S. 1980. *The Economics of Village Bali: Three Perspectives.* New Delhi: The Ford Foundation.

Popkin, B.M. 1993. Nutritional patterns and transitions. *Population and Development Review* 19(1): 138–157.

Popp, A., Lotze-Campen, H. and Bodirsky, B. 2010. Food consumption, diet shifts and associated non-CO2 greenhouse gases from agricultural production. *Global Environmental Change* 20(3): 451–462.

Popp, J., Pető, K. and Nagy, J. 2013. Pesticide productivity and food security. A review. *Agronomy for Sustainable Development* 33(1): 243–255.

Posey, D. 1985. Indigenous management of tropical forest ecosystems. *Agroforestry Systems* 3: 139–158.

POST. 2017. Environmentally Sustainable Agriculture. Parliamentary Office of Science and Technology, No. 557. London.

Pradhan, P. 2000. Farmer managed irrigation systems in Nepal at the crossroad. Paper presented at the 8th Biennial Conference of the International Association for the Study of Common Property (IASCP), Bloomingdale, Indiana. URL: http://dlc.dlib.indiana.edu/dlc/bitstream/handle/10535/331/pradhanp041500.pdf?sequence=1

Praneetvatakul, S. and Waibel, H. 2006. Impact assessment of farmer field schools using a multi-period panel data model. International Association of Agricultural Economists Conference, Gold Coast, Australia, 12–18 August.

Praneetvatakul, S., Schreinemachers, P., Pananurak, P. and Tipraqsa, P. 2013. Pesticides, external costs and policy options for Thai agriculture. *Environmental Science & Policy* 27: 103–113.

Pretty, J. 1991. Farmers' extension practice and technology adaptation: agricultural revolution in 17–19th century Britain. *Agriculture and Human Values* 8: 132–148.

Pretty, J. 1995. Participatory learning for sustainable agriculture. *World Development* 23(8): 1247–1263.

Pretty, J. 1997. The sustainable intensification of agriculture. *Natural Resources Forum* 21(4): 247–256.

Pretty, J. 2002. *Agri-Culture. Reconnecting People, Land and Nature.* London: Earthscan.

Pretty, J. 2003. Social capital and the collective management of resources. *Science* 302: 1912–1915.

Pretty, J.N. (ed.) 2005. *The Pesticide Detox: Towards a More Sustainable Agriculture.* London: Earthscan.

Pretty, J. 2008. Agricultural sustainability: concepts, principles and evidence. *Philosophical Transactions of the Royal Society of London B* 363(1491): 447–466.

Pretty, J. 2013. The consumption of a finite planet: well-being, convergence, divergence and the nascent green economy. *Environmental and Resource Economics* doi: 10.1007/s10640-013-9680-9

Pretty, J.N. 2014. *The Edge of Extinction.* Ithaca, NY: Cornell University Press.

Pretty, J. 2015. Blog post. 4 July. www.julespretty.com/no-33-the-contrary-farmers/

Pretty, J.N. and Frank, B. 2000. Participation and social capital formation in natural resource management: achievements and lessons. International Landcare Conference, Melbourne.

Pretty, J. and Ward, H. 2001. Social capital and the environment. *World Development* 29(2): 209–227.

Pretty, J.N. and Waibel, H. 2005. Paying the price: the full cost of pesticides. In: Pretty, J. (ed.) *The Pesticide Detox: Towards a More Sustainable Agriculture.* London: Earthscan.

Pretty, J.N. and Bharucha, Z.P. 2014. Sustainable intensification in agricultural systems. *Annals of Botany* 205: 1–26.

Pretty, J. and Bharucha, Z.P. 2015. Integrated pest management for sustainable intensification of agriculture in Asia and Africa. *Insects* 6: 152–182.

Pretty, J.N., Ball, A.S., Lang, T. and Morison, J.I. 2005. Farm costs and food miles: an assessment of the full cost of the UK weekly food basket. *Food Policy* 30(1): 1–19.

Pretty, J.N., Noble, A.D., Bossio, D., Dickson, J., Hine, R.E., Penning de Vries, F.W.T. and Morrison, J.I.L. 2006. Resource-conserving agriculture increases yields in developing countries. *Environmental Science and Technology* 40: 1114–1119.

Pretty, J., Toulmin, C. and Williams, S. (eds) 2011a. Sustainable intensification in African agriculture. *International Journal of Agricultural Sustainability* 9(1): 1–241.

Pretty, J., Toulmin, C. and Williams, S. 2011b. Sustainable intensification in African agriculture. *International Journal of Agricultural Sustainability* 9(1): 5–24.

Pretty, J.N., Bharucha, Z.P., Hama Garba, M., Midega, C., Nkonya, E., Settle, W. and Zingore, S. 2014. Foresight and African agriculture: innovations and policy opportunities. Report to the UK Government Foresight Project. https://www.gov.uk/government/uploads/system/uploads/attachment_data/file/300277/14-533-future-african-agriculture.pdf

Pretty, J., Barton, J., Bharucha, Z.P., Bragg, R., Pencheon, D., Wood, C. and Depledge, M.H. 2015. Improving health and well-being independently of GDP: dividends of greener and prosocial economies. *International Journal of Environmental Health Research* 11: 1–26.

Proshika. 2017. URL www.proshika.org

Qaim, M. 1999. The Economic Effects of Genetically Modified Orphan Commodities: Projections for Sweet potato in Kenya. ISAA Briefs No. 13. Ithaca, NY: ISAA; and Bonn: ZEF. URL: www.isaaa.org/resources/publications/briefs/13/download/isaaa-brief-13-1999.pdf

Raintree, J.B. and Warner, K. 1986. Agroforestry pathways for the intensification of shifting cultivation. *Agroforestry Systems* 4: 39–54.

Ramisch, J.J., Misiko, M.T., Ekise, I.E. and Mukalama, J.B. 2006. Strengthening 'folk ecology': community-based learning for integrated soil fertility management, western Kenya. *International Journal of Agricultural Sustainability* 4(2): 154–168.

Raney, T. 2006. Economic impact of transgenic crops in developing countries. *Current Opinion in Biotechnology* 17(2): 174–178.

Rao, A.N., Johnson, D.E., Sivaprasad, B., Ladha, J.K. and Mortimer, A.M. 2007. Weed management in direct-seeded rice. *Advances in Agronomy* 93: 153–255.

Rao, I., Peters, M., van der Hoek, R., Castro, A., Subbarao, G., Cadisch, G. and Rincón, A. 2014. Tropical forage-based systems for climate-smart livestock production in Latin America. *Rural 21* 04/2014: 12–15.

Ravindranath, N.H. and Sudha, P. 2004. *Joint Forest Management in India: Spread, Performance and Impact.* Hyderabad: Universities Press.

Raworth, K. 2017. *Doughnut Economics: Seven Ways to Think Like a 21st-Century Economist.* London: Chelsea Green Publishing.

Reaganold, J.P. and Wachter, J.M. 2016. Organic agriculture in the 21st century. *Nature Plants* 2(2): 15221.

Reddy, V.R. and Reddy, P. 2005. How participatory is participatory irrigation management? Water users' associations in Andhra Pradesh. *Economic & Political Weekly* 40(53): 5587–5595.

Reij, C.P. and Smaling, E.M.A. 2008. Analyzing successes in agriculture and land management in sub-Saharan Africa: Is macro-level gloom obscuring positive micro-level change? *Land Use Policy* 25: 410–420.

Reij, C., Tappan, G. and Smale, M. 2009. Agroenvironmental transformation in the Sahel: another kind of 'green revolution'. IFPRI Discussion Paper 00914. Washington, DC: International Food Policy Research Institute.

Renewables 21. 2017. *Global Status Report 2017*. Renewable Energy Policy Network for the 21st Century. Paris: REN21.

Resende, Á.V., Furtini Neto, A.E., Alves, V.M.C., Curi, N., Muniz, J.A., Faquin, V. and Kinpara, D.I. 2007. Phosphate efficiency for corn following *Brachiaria* grass pasture in the Cerrado Region. *Better Crops* 91(1): 17–19.

Rockström, J., Steffen, W., Noone, K., Persson, Å., Chapin, F.S., Lambin, E.F., *et al.* 2009. A safe operating space for humanity. *Nature* 461(7263): 472–475.

Rockström, J., Williams, J., Daily, G., Noble, A., Matthews, N., Gordon, L., *et al.* 2017. Sustainable intensification of agriculture for human prosperity and global sustainability. *Ambio* 46: 4–17.

Rogelj, J., Hare, B., Nabel, J., Macey, K., Schaeffer, M., Markmann, K. and Meinshausen, M. 2009. Halfway to Copenhagen, no way to 2 °C. *Nature Reports Climate Change*, pp. 81–83.

Röling, N. 1996. Towards an interactive agricultural science. *European Journal of Agricultural Education and Extension* 2: 3–48.

Röling, N. 2010. The impact of agricultural research: evidence from West Africa. *Development in Practice* 20(8): 959–971.

Röling, N.G. and Wagemakers. M.A.E. (eds) 1997. *Facilitating Sustainable Agriculture*. Cambridge: Cambridge University Press.

Rosenstock, T.S., Tully, K.T., Arias-Navarro, C., Neufeldt, H., Butterbach-Bahl, K. and Verchot, L.V. 2014. Agroforestry with N2-fixing trees: sustainable development's friend or foe? *Current Opinion in Environmental Sustainability* 6: 15–21.

Rosset, P.M. and Martínez-Torres, M.E. 2012. Rural social movements and agroecology: context, theory, and process. *Ecology and Society* 17(3): 17.

Rosset, P.M., Machín Sosa, B., Roque Jaime, A.M. and Ávila Lozano, D.R. 2011. The Campesino-to-Campesino agroecology movement of ANAP in Cuba: social process methodology in the construction of sustainable peasant agriculture and food sovereignty. *Journal of Peasant Studies* 38(1): 161–191.

Rowe, W.C. 2009. Kitchen gardens in Tajikistan: the economic and cultural importance of small-scale private property in a post-Soviet society. *Human Ecology* 37: 691–703.

Royal Society. 2009. *Reaping the Benefits: Science and the Sustainable Intensification of Global Agriculture*. London: Royal Society.

Royal Society. 2012. *People and the Planet*. London: Royal Society.

Sahota, A. 2016. The global market for organic food and drink. In: Willer, H. and Lernoud, J. (eds) *The World of Organic Agriculture. Statistics and Emerging Trends* (pp.133–137). Frick, Switzerland: Research Institute of Organic Agriculture (FiBL); and Bonn: IFOAM – Organics International.

Sanyang, S., Taonda, S.J.B., Kuiseu, J., Coulibaly, N.T. and Konaté, L. 2016. A paradigm shift in African agricultural research for development: the role of innovation platforms. *International Journal of Agricultural Sustainability* 14(2): 187–213.

Sawadogo, H. 2011. Rehabilitation of degraded lands by using soil and water conservation techniques in the north western region of Burkina Faso. *International Journal of Agricultural Sustainability* 9(1): 120–128.

Schmitz, P.M. 2001. Overview of cost–benefit assessment. In OECD workshop on the Economics of Pesticide Risk Reduction in Agriculture, Copenhagen, 28–30 November. Paris: OECD.

Scopel, E., Triomphe, B., dos Santos Ribeiro, Mde F., Séguy, L., Denardin, J.E. and Kochhann, R.A. 2004. Direct seeding mulch-based cropping systems (DMC) in Latin America. In: Fischer, R.A. (ed.) *New Directions for a Diverse Planet*. Proceedings of the 4th International Crop Science Congress. Brisbane, Australia.

Sendzimir, J.C.P., Reij, C. and Magnuszewski, P. 2011. Rebuilding resilience in the Sahel: regreening in the Maradi and Zinder regions of Niger. *Ecology and Society* 16(3): 1. Available online: http://dx.doi.org/10.5751/ES-04198-160301

SERIO (2012) The Value of the Community Food Sector – An Economic Baseline of Community Food Enterprises. Plymouth University.

Settle, W. and Hama Garba, M. 2011. Sustainable crop production intensification in the Senegal and Niger River Basins of francophone West Africa. *International Journal of Agricultural Sustainability* 9(1): 171–185.

Settle, W., Soumaré, M., Sarr, M., Hama Garba, M. and Poisot, A. 2014. Reducing pesticide risks to farming communities: cotton farmer field schools in Mali. *Philosophical Transactions of the Royal Society (B)* 369: 20120277. http://dx.doi.org/10.1098/rstb.2012.0277

Seufert, V., Ramankutty, N. and Foley, J.A. 2012. Comparing the yields of organic and conventional agriculture. *Nature* 485: 229–232.

Shaner, D.L. 2014. lessons learned from the history of herbicide resistance. *Weed Science* 62(2): 427–431.

Sheehy, J.E., Peng, S., Dobermann, A., Mitchell, P.L., Ferrer, A., Yang, J., *et al.* 2004. Fantastic yields in the system of rice intensification: fact or fallacy? *Field Crops Research* 88(1): 1–8.

Sheehy, J.E., Sinclair, T.R. and Cassman, K.G. 2005. Curiosities, nonsense, non-science and SRI. *Field Crops Research* 91: 355–356.

Shetty, P. 2012. Public health: India's diabetes time bomb. *Nature* 485: S14–16 doi: 10.1038/485S14a

Sileshi, G.W., Debusho, L.K. and Akinnifesi, F.K. 2012. Can integration of legume trees increase yield stability in rainfed maize cropping systems in southern Africa? *Agronomy Journal* 104(5): 1392–1398.

Silici, L., Ndabe, P., Friedrich, T. and Kassam, A. 2011. Harnessing sustainability, resilience and productivity through conservation agriculture: the case of likoti in Lesotho. *International Journal of Agricultural Sustainability* 9(1): 137–144.

Sinha, P. 2014. Status of Participatory Irrigation Management (PIM) in India. National Convention of Presidents of Water Users Associations, Ministry of Water Resources India NPIM, New Delhi, 7–8 November.

Smith, A., Snapp, S., Chikowo, R., Thorne, P., Bekunda, M. and Glover, J. 2017. Measuring sustainable intensification in smallholder agroecosystems: a review. *Global Food Security* 12: 127–138.

Smith, P. 2013. Delivering food security without increasing pressure on land. *Global Food Security* 2: 18–23.

Smith, P., Martino, D., Cai, Z., Gwary, D., Janzen, H., Kumar, P., *et al.* 2008. Greenhouse gas mitigation in agriculture. *Philosophical Transactions of the Royal Society of London B: Biological Sciences* 363(1492): 789–813.

Smith-Spangler, C., Brandeau, M.L., Hunter, G.E., Bavinger, J.C., Pearson, M., Eschbach, P.J., *et al.* 2012. Are organic foods safer or healthier than conventional alternatives? A systematic review. *Annals of Internal Medicine* 157(5): 348–366.

Smoyer-Tomic, K.E., Spence, J.C., Raine, K.D., Amrhein, C., Cameron, N., Yasenovskiy, V., *et al.* 2008. The association between neighborhood socioeconomic status and exposure to supermarkets and fast food outlets. *Health & Place* 14(4): 740–754.

Snapp, S.S., Blackie, M.J., Gilbert, R.A., Bezner-Kerr, R. and Kanyama-Phiri, G.Y. 2010. Biodiversity can support a greener revolution in Africa. *PNAS* 107(48): 20840–20845.

Solh, M. and van Ginkel, M. 2014. Drought preparedness and drought mitigation in the developing world's drylands. *Weather and Climate Extremes* 3: 62–66.

Sorrell, S. 2007. The Rebound Effect: An Assessment of the Evidence for Economy-wide Energy Savings from Improved Energy Efficiency. UK Energy Research Centre.

Sorrenson, W.J. 1997. Financial and economic implications of no-tillage and crop rotations compared to conventional cropping systems. TCI Occasional Paper Series No. 9. Rome: FAO.

Spargo, J.T., Alley, M.M., Follet, R.F. and Wallace, J.V. 2008. Soil carbon sequestration with continuous no-till management of grain cropping systems in the Virginia coastal plain. *Soil and Tillage Research* 100(1–2): 13–140.

Sponsel, L.E. 1989. Farming and foraging: a necessary complementarity in Amazonia? In: Kent, S. (ed.) *Farmers as Hunters: The Implications of Sedentism*. Cambridge: Cambridge University Press.

Steffen, W., Richardson, K., Rockström, J., Cornell, S.E., Fetzer, I., Bennett, E.M., *et al.* 2015. Planetary boundaries: guiding human development on a changing planet. *Science* 347(6223): 1259855.

Stern, N. 2007. *The Economics of Climate Change: The Stern Review*. Cambridge: Cambridge University Press.

Stern, N. and Rydge, J. 2012. The new energy–industrial revolution and an international agreement on climate change. *Economics of Energy & Environmental Policy* 1: 1–19.

Stevens, G.A., Singh, G.M., Lu, Y., Danaei, G., Lin, J.K., Finucane, M.M., *et al.* 2012. National, regional, and global trends in adult overweight and obesity prevalences. *Population Health Metrics* 10: 22.

Stevenson, J.R., Serraj, R. and Cassman, K.G. 2014. Evaluating conservation agriculture for small-scale farmers in sub-Saharan Africa and South Asia. *Agriculture, Ecosystems and Environment* 187: 1–10.

Stoop, W. 2011. The scientific case for system of rice intensification and its relevance for sustainable crop intensification. *International Journal of Agricultural Sustainability* 9(3): 443–455.

Stoop, W., Uphoff, N. and Kassam, A. 2002. A review of agricultural research issues raised by the system of rice intensification (SRI) from Madagascar: opportunities for improving farming systems for resource-poor farmers. *Agricultural Systems* 71(3): 249–274.

Styger, E., Aboubacrine, G., Attaher, M.A. and Uphoff, N. 2011. The system of rice intensification as a sustainable agricultural innovation: introducing, adapting and scaling up a system of rice intensification practices in the Timbuktu region of Mali. *International Journal of Agricultural Sustainability* 9(1): 67–75.

Sumberg, J., Thompson, J. and Woodhouse, P. 2013. Why agronomy in the developing world has become contentious. *Agriculture and Human Values* 30(1): 71–83.

Sustainable Food Trust. 2017. The hidden cost of UK food. Available online: http://sustainablefoodtrust.org/wp-content/uploads/2013/04/HCOF-Report-online-version.pdf

Swaminathan, M.S. 1989. Agricultural production and food security in Africa. In: d'Orville, H. (ed.) *The Challenges of Agricultural Production and Food Security in Africa*. A report of the proceedings of an international conference organized by the Africa Leadership Forum 27–30 July. Ota, Nigeria. www.africaleadership.org/rc/the%20challenges%20of%20agricultural.pdf#page=23 (31 August 2014).

Swaminathan, M.S. 2000. An evergreen revolution. *Biologist* 47(2): 85–89.

Szuter, C.R. and Bayham, F.E. 1989. Sedentism and prehistoric animal procurement among desert horticulturalists of the North American Southwest. In: Kent, S. (ed.) *Farmers as Hunters: The Implications of Sedentism*. Cambridge: Cambridge University Press.

TakePart. 2016. Organic Farming in the US is now bigger than ever. URL: www.takepart.com/article/2016/11/10/organic-crop-acreage/

Taylor, J.R. and Lovell, S.T. 2012. Urban home food gardens in the Global North: research traditions and future directions. *Agriculture and Human Values* 31(2): 285–305.

TEEB–Agrifood, 2017. TEEB–Agrifood Foundations Wireframe. URL: www.teebweb.org/agriculture-and-food/foundations-wireframe/

Thakur, A.K. and Uphoff, N.T. 2017. How the system of rice intensification can contribute to climate-smart agriculture. *Agronomy Journal* 109(4): 1163–1182.

Thakur, A.K., Uphoff, N. and Antony, E. 2010a. An assessment of physiological effects of system of rice intensification (SRI) practices compared with recommended rice cultivation practices in India. *Experimental Agriculture* 46(1): 77–98.

Thakur, A.K., Rath, S., Roychowdhury, S. and Uphoff, N. 2010b. Comparative performance of rice with system of rice intensification (SRI) and conventional management using different plant spacings. *Journal of Agronomy and Crop Science* 196: 146–159.

Thakur, A.K., Rath, S., Patil, D.U. and Kumar, A. 2011. Effects on rice plant morphology and physiology of water and associated management practices of the system of rice intensification and their implications for crop performance. *Paddy and Water Environment* 9: 13–24.

Thakur, A.K., Rath, S. and Mandal, K.G. 2013. Differential responses of system of rice intensification (SRI) and conventional flooded-rice management methods to applications of nitrogen fertilizer. *Plant and Soil* 370: 59–71.

Thakur, A.K., Kassam, A.H, Stoop, W.A. and Uphoff, N.T. 2016. Scientific underpinnings of the system of rice intensification (SRI): what is known so far. *Advances in Agronomy* 135: 147–178.

The National Allotment Society. 2014. Brief History of Allotments. www.nsalg.org.uk/allotment-info/brief-history-of-allotments/

Thompson, B. and Amoroso, L. 2011. FAO's Approach to Nutrition-Sensitive Agricultural Development. Rome: FAO www.fao.org/fileadmin/user_upload/agn/pdf/FAO_Approach_to_Nutrition_sensitive_agricultural_development.pdf

Thomson, A.J., Giannopoulos, G., Pretty, J.N., Braggs, E.M. and Richardson, D.J. 2012. Biological sources and sinks of nitrous oxide and strategies to mitigate emissions. *Philosophical Transactions of the Royal Society B* 367(1593): 1157–1168.

Tilman, D., Balzer, C., Hill, J. and Befort, B.L. 2011. Global food demand and the sustainable intensification of agriculture. *PNAS* 108: 20260–20264.

Tilman, E., Cassman, K.G., Matson, P.A., Naylor, R. and Polasky, S. 2002. Agricultural sustainability and intensive production practices. *Nature* 418: 617–677.

Tomekpe, K. and Ganry, J. 2011. CARBAP and innovation on the plantain banana in West and Central Africa. *International Journal of Agricultural Sustainability* 9(1): 264–273.

Tomlinson, I. 2013. Doubling food production to feed the 9 billion: a critical perspective on a key discourse of food security in the UK. *Journal of Rural Studies* 29: 81–90.

Traore, A. and Bickersteth, S. 2011. Addressing the challenges of agricultural service provision: the case of Oxfam's Strategic Cotton Programme in Mali. *International Journal of Agricultural Sustainability* 9(1): 82–90.

Tripp, R., Wijeratne, M. and Piyadasa, V.H. 2005. What should we expect from farmer field schools? A Sri Lanka case study. *World Development* 33: 1705–1720.

Truog, E. 1946. 'Organics only?' Bunkum! *The Land* 5: 317–321.

Tubiello, F.N., Salvatore, M., Rossi, S., Ferrara, A., Fitton, N. and Smith, P. 2013. The FAOSTAT database of greenhouse gas emissions from agriculture. *Environmental Research Letters* 8(1): 015009.

Tuck, S.L., Winqvist, C., Mota, F., Ahnström, J., Turnbull, L.A. and Bengtsson, J. 2014. Land-use intensity and the effects of organic farming on biodiversity: a hierarchical meta-analysis. *Journal of Applied Ecology* 51(3): 746–755.

Turmel, M., Turner, B. and Whalen, J.K. 2011. Soil fertility and the yield response to the system of rice intensification. *Renewable Agriculture and Food Systems* 26(3): 185–192.

UNCSD. 2012. United Nations Conference on Sustainable Development. At www.uncsd2012.org/rio20/

Undersander, D., Albert, B., Cosgrove, D., Johnson, D. and Peterson, P. 2002. *Pastures for Profit: A Guide to Rotational Grazing.* Cooperative Extension Publishing. University of Wisconsin-Extension.

UNEP. 2011. Towards a Green Economy: Pathways to Sustainable Development and Eradication of Poverty. Nairobi: UNEP.

UNEP. 2014. The Emissions Gap Report. A UN Synthesis Report. Nairobi: UNEP.

UNICEF. 2012. Levels and Trends in Child Mortality. New York: UNICEF. Uphoff, N. 1999. Agroecological implications of the system of rice intensification (SRI) in Madagascar. *Environment, Development and Sustainability* 1(3–4): 297–313.

Uphoff, N. 2003. Higher yields with fewer external inputs? The system of rice intensification and potential contributions to agricultural sustainability. *International Journal of Agricultural Sustainability* 1(1): 38–50.

Uphoff, N. 2015. The System of Rice Intensification: Responses to Frequently-Asked Questions. Ithaca, NY: CreateSpace. http://sri.cals.cornell.edu/aboutsri/SRI_FAQs_Uphoff_2016.pdf [accessed 8 May 2017].

Uphoff, N., Kassam, A. and Stoop, W. 2008. A critical assessment of a desk study comparing crop production systems: the example of the 'system of rice intensification' vs. 'best management practice'. *Field Crops Research* 108(1): 109–114.

USDA. 1955. Agricultural Statistics. National Agricultural Statistics Service (NASS). www.NASS.usda.gov/publications

USDA. 2012. Agricultural Statistics. National Agricultural Statistics Service (NASS). www.NASS.usda.gov/publications

USDA. 2014. Agricultural Statistics. National Agricultural Statistics Service (NASS). Washington, DC.

USDA. 2016. Farmers' Market Promotion Program. 2016 Report. Connecting Rural and Urban Communities. URL: https://www.ams.usda.gov/reports/farmers-market-promotion-program-2016-report

USDA. 2017. Certified Organic Survey. 2016 Summary. USDA. URL: http://usda.mannlib.cornell.edu/usda/current/OrganicProduction/OrganicProduction-09-20-2017_correction.pdf

Uysal, Ö.K. and Atış, E. 2010. Assessing the performance of participatory irrigation management over time: a case study from Turkey. *Agricultural Water Management* 97(7): 1017–1025.

van den Berg, J. 2004. *IPM Farmer Field Schools: A Synthesis of 25 Impact Evaluations.* Rome: Global IPM Facility.

van den Berg, H. and Jiggins, J. 2007. Investing in farmers – the impacts of farmer field schools in relation to integrated pest management. *World Development* 35(4): 663–686.

Vickers, W.T. 1989. Patterns of foraging and gardening in a semi-sedentary Amazonian community. In: Kent, S. (ed.) *Farmers as Hunters: The Implications of Sedentism.* Cambridge: Cambridge University Press.

Vogl, C.R. and Vogl-Lukasser, B. 2003. Tradition, dynamics and sustainability of plant species composition and management in homegardens on organic and non-organic small scale farms in Alpine Eastern Tyrol, Austria. *Biological Agriculture and Horticulture* 21: 149–166.

Wang, J., Huang, J., Zhang, L., Huang, Q. and Rozelle, S. 2010. Water governance and water use efficiency: the five principles of WUA management and performance in China. *Journal of the American Water Resources Association* 46(4): 665–685.

Wang, X.B., Oenema, O., Hoogmoed, W.B., Perdok, U.D. and Cai, D.X. 2006. Dust storm erosion and its impact on soil carbon and nitrogen losses in northern China. *Catena* 66: 221–227

Ward, B. and Dubos, R. 1972. *The Care and Maintenance of a Small Planet.* Harmondsworth: Penguin.

WCED (World Commission on Environment and Development). 1987. *Our Common Future.* (Brundtland Report). Oxford: Oxford University Press.

Weerakoon, W.M.W., Mutunayake, M.M.P., Bandara, C., Rao, A.N., Bhandari, D.C. and Ladha, J.K. 2011. Direct-seeded rice culture in Sri Lanka: lessons from farmers. *Field Crops Research* 121(1): 53–63.

Wennink, B. and Heemskerk, W. 2004. Building social capital for agricultural innovation: experiences with farmer groups in sub-Saharan Africa. Bulletin 368. Amsterdam: Royal Tropical Institute (KIT). URL: www.iscom.nl/publicaties/buildingsocialcapitalagriculture.pdf

Wibbelmann, U., Schmutz, J., Wright, D., Udall, F., Rayns, M., Kneafsey, L., *et al.* 2013. Mainstreaming agroecology: implications for global food and farming systems. Centre for Agroecology and Food Security Discussion Paper. Coventry: Centre for Agroecology and Food Security.

Wibberley, G.P. 1959. *Agriculture and Urban Growth: A Study of the Competition for Rural Land.* London: Michael Joseph.

Wicke, B., Sikkema, R., Dornburg, V. and Faaij, A. 2011. Exploring land use changes and the role of palm oil production in Indonesia and Malaysia. *Land Use Policy* 28(1): 193–206.

Willer, H. and Lernoud, J. (eds) 2016. *The World of Organic Agriculture. Statistics and Emerging Trends 2016* (pp. 1–336). Frick, Switzerland: Research Institute of Organic Agriculture (FiBL); and Bonn: IFOAM – Organics International.

Williamson, S., Ball, A.S. and Pretty, J.N. 2008. Trends for pesticide use and safer pest management in four African countries. *Crop Protection* 27: 1327–1334.

WinklerPrins, A.M.G.A. 2003. House-lot gardens in Santarem, Pará, Brazil: linking rural with urban. *Urban Ecosystems* 6(1–2): 43–65.

WinklerPrins, A.M.G.A. and de Souza, P. 2005. Surviving the city: urban home gardens and the economy of affection in the Brazilian Amazon. *Journal of Latin American Geography* 4(1): 107–126.

World Bank. 2012. *Inclusive Green Growth. The Pathway to Sustainable Development.* Washington, DC: World Bank.

Worster, D. 1994. *Nature's Economy: A History of Ecological Ideas.* Cambridge: Cambridge University Press.

WRG (2030 Water Resources Group). 2009. Charting Our Water Future. Economic frameworks to inform decision-making. Available online: www.mckinsey.com/client_service/sustainability/latest_thinking/charting_our_water_future

Wright, D., Camden-Pratt, C. and Hill, S. (eds) 2011. *Social Ecology: Applying Ecological Understanding to Our Lives and Our Planet.* Stroud: Hawthorn Press.

WWF. 2010. *Living Planet Report.* London and Oakland: WWF and Global Footprint Network.

Xinhua. 2016. CPC and State Council Guide Opinion on Using New Development Concepts to Accelerate Agricultural Modernisation and Realise Moderate Prosperity Society. Available online: http://news.xinhuanet.com/fortune/2016-01/27/c_1117916568.htm

Yang, P.Y., Zhao, Z.H. and Shen, Z.R. 2014. Experiences with implementation and adoption of integrated pest management in China. In: Peshin, R. and Pimentel, D. (eds) *Integrated Pest Management* (pp. 307–330). Dordrecht: Springer.

Young, M. and Esau, C. (eds) 2016. *Transformational Change in Environmental and Natural Resource Management: Guidelines for Policy Excellence.* London: Earthscan.

Yu, Y.X., Zhao, C.Y. and Jia, H.T. 2015. Ability of split urea applications to reduce nitrous oxide emissions: a laboratory incubation experiment. *Applied Soil Ecology* 100: 75–80. doi: 10.1016/j.apsoil.2015.12.009

Zhang, W.J., Jiang, F.B. and Ou, J.F. 2011. Global pesticide consumption and pollution: with China as a focus. *Proceedings of the International Academy of Ecology and Environmental Sciences* 1(2): 125–144.

Zhang, W., Cao, G., Li, X., Zhang, H., Wang, C., Liu, Q., *et al.* 2016. Closing yield gaps in China by empowering smallholder farmers. *Nature* 537(7622): 671–674.

Zhao, L.M., Wu, L.H., Li, Y.S., Lu, X.H., Zhu, D.F. and Uphoff, N. 2009. Influence of the system of rice intensification on rice yield and nitrogen and water use efficiency with different application rates. *Experimental Agriculture* 45: 275–286.

Zomer, R.J., Trabucco, A., Coe, R. and Place, F. 2009. Trees on Farm: Analysis of Global Extent and Geographical Patterns of Agroforestry. ICRAF Working Paper no. 89. Nairobi: World Agroforestry Centre.

INDEX

Abubakr, D.M. 114
additive systems 47
affluence 2, 11, 14, 25, 27, 29, 31
Afghanistan 58, 80–1
Africa 3, 13, 16–17, 20, 22; and agroforestry
70; and aquaculture 74; and CA 66;
and dryland greening 82–3; and
farmer field schools 117; and food
24, 26; and industrialised countries
93, 99; and integrated systems 80; and
patch intensification 72; and policy
frameworks 138–9; and policy-making
137; and push-pull systems 117; and
redesign 127, 145–6, 150; and SI 35, 39,
41, 43, 49–50, 52, 58; and small farms
60, 62, 64, 66, 69–70; and SRI/SCI 80
African Research Centre on Banana and
Plantain (CAR-BAP) 138
African Union 67
Africare 79, 81
Agreco Farmers' Organisation 129–30
agri-environment scheme (AES) 90
agriculture 1–6, 7–10; agricultural
extension 112–15, 118; agricultural
revolutions 1, 10–16; agricultural
science/scientists 62, 67–8, 72, 80, 94,
128, 150; agriculture biologique 97;
agriculture durable 3, 50; agrochemicals
20, 63, 75; agroeconomics 3, 47, 62;
in developing countries 45–58; and
effectiveness of SI 45–58; and greener
economies 132–41; in industrialised
countries 45–58; and knowledge
economy 107–31; and redesign 107–31,
142–51; and SI 31–44; and small farms

45–58; and social capital 107–31; in
twenty-first century 7–30; and
world-building 142–51
Agro-Pastoral Field Schools (APFS) 117,
147
agroecology 2, 5, 32–3, 35, 37–8; and
effectiveness of SI 45–7, 52; and farmer
field schools 114–15; and greener
economies 135; in industrialised
countries 88; and participation 112; and
policy frameworks 137; and redesign 91,
127–8, 146, 150; and SI support 141;
and small farms 64–6; and SRI/SCI
76–7, 79–80
agroecosystems 5–6, 32–3, 41, 45, 49;
agroecosystem analysis 115; and CA
67–8; and effectiveness of SI 52, 54;
and farmer field schools 116; and
greener economies 135; in industrialised
countries 88; and integrated systems
80; and participation 112; and policy-
making 136; and redesign 150; and
slash-and-mulch techniques 84; and
small farms 63, 66
agroforestry 39, 46, 50–1, 56, 69–71; and
dryland greening 83; and fertilisers
69–71; and greener economies 135; and
participation 111; and policy-making
135; and redesign 111, 129, 144, 147;
and slash-and-mulch techniques 85; and
small farms 82–3, 85, 87; and wisdom
networks 129; and world-building 144
agronomics/agronomists 76, 88, 105, 118,
135, 137, 150
Alexandra, J. 105

Printed in the United States
by Baker & Taylor Publisher Services